THE FULL STORY OF THE ANGLO-FRENCH SST

CONCORDE

". . . new developments in the field of air transportation can, in the hands of men of good will, provide the one instrument that can forever eliminate the barriers of time and distance and, as a consequence, the barriers of ignorance and misunderstanding, which, since the dawn of civilization, have set men apart and against one another . . ."

—Eddie Rickenbacker

Edward V. Rickenbacker, RICKENBACKER.
(Englewood Cliffs, N.J., Prentice-Hall, Inc., 1967)

Concorde's immortal beauty is clearly evident in this
scene with British Airway's new paint scheme.
BRITISH AIRWAYS PHOTO

CONCORDE

THE FULL STORY OF THE ANGLO~FRENCH SST

by Richard K. Schrader

PICTORIAL HISTORIES PUBLISHING COMPANY
MISSOULA, MONTANA

LIBRARY OF CONGRESS
CATALOG CARD NUMBER 89-60233

ISBN 0-929521-16-1

First Printing: April 1989

PRINTED IN THE UNITED STATES OF AMERICA

Typography: Arrow Graphics & Typography
Cover Graphics: Kirk Johnson
Layout: Stan Cohen

About the Author

The author is a commercial pilot and has been flying since 1966. Although his special interest is in vintage aircraft, and he has logged many hours in biplanes across the country, he was caught up in the Concorde as a result of the engineering and devotion to safety that went into her development. "Of course, she is also a very beautiful airplane," he adds. He felt compelled to provide an account of the Anglo-French SST from an American's point of view, and more than three years of labor went into the project. By necessity, much of the research and photographic material had to be obtained from Europe.

Mr. Schrader also has an avid interest in the history of Airborne troops and has written many articles on the subject. He completed U.S. Army paratrooper training at Fort Benning and went on to jump with the elite 82nd Airborne Division. Presently he is a contributing writer for *Air Classics*, a variety of other magazines and a large newspaper. He hails from Newark, New Jersey, and is a graduate of Newark State College.

PICTORIAL HISTORIES PUBLISHING COMPANY
713 South Third West
Missoula, Montana 59801

Contents

Acknowledgments

Completion of this book would have been impossible without the help of many people and organizations. I wish to express my appreciation to those who gave their time to provide me with historical and technical information.

Edward Tourtellotte, Gail Muntner, Bruce Haxthausen, Ariane Scarpa and Tina Pogach of Air France never tired of my questions and responded with nothing but kindness. Heartfelt thanks also is due Captain Michel Butel, who provided much more than a close look at the supersonic transport and the flight deck. The feeling in his voice when speaking of Concorde revealed the intense pride of the crews who fly her. In addition, Jacqueline Didelot and her well-mannered flight attendants displayed professionalism and hospitality in showing the cabin.

Joyce Brown, Douglas Nelms, Tony Lewis, Nigel Oldfield, M.V. Brown, Sean Green, George Dahlman and Leslie Nelson of British Aerospace were superb sources of information. Nothing at all seemed to be too much trouble; they contributed detailed material on Concorde and current BAe projects with civility.

André Turcat, former chief test pilot for Sud Aviation and Aerospatiale, granted a personal interview. He vividly conveyed the spirit of the Anglo-French SST enterprise and reconstructed his thoughts and feelings of the maiden flight. He also took time to provide historic photographs.

Deborah Bernstein of British Airways supplied material with warmth and dedication to the project, which more than made up for the indifference of her predecessors. Her enthusiasm for Concorde clearly abounded.

John Wesler, director of Environment and Energy in the Federal Aviation Administration, provided a copy of the official and complete transcript of the Washington, D.C., hearing on Concorde and of the Transportation secretary's historic decision on that matter. The several hundred pages of testimony and judgment enabled excellent insight into the Americans' viewpoint on the SST.

Nick Komons, the FAA's agency historian, went out of his way to locate additional information on the supersonic subject.

Additional airline personnel helped with a supply of relevant facts and photographs: James Arey and Ann Whyte of Pan American World Airways; R.C. Novak of United Airlines; Odette Fodor of KLM, Royal Dutch Airlines; and the public-relations crews of American, Delta, and Trans World Airlines.

Assistance and photographs also were provided by André Fleury of the Académie Nationale de l'Air et de l'Espace; Gabrielle le Mire-Cahn of Aerospatiale; Loretta M. Kulaga of CASA; Angel Biasatti and Marilyn Brown of the Dallas/Fort Worth International Airport Public Information Department; Robert Hawk of Fokker Aircraft USA; Russ Harbourne of the Intercontinental Dynamics Corporation; Mrs. M.M. Chapman of the Royal Aircraft Establishment; Mike Savage of Saab Aircraft International; Ronald G. Crawford of Short Brothers; and the public-relations staffs of Airbus Industrie of North America, General Dynamics, Lockheed and McDonnell-Douglas.

To the people of France and Britain,
who met the challenge.

To the American soldiers, who made it all possible.

And to Mr. William Coleman, who embodied the
spirit of fair play.

Chapter One:
Historical Background

Concorde is the epitome of elegance. The flow of her sleek fuselage and graceful wing illustrates refinement and reveals the reason for her great speed. Her unique vertical stabilizer and rudder—sweeping back, then forward—present a finishing touch that caps the aesthetic value of her being. Yet, something besides her beauty is readily apparent. Sitting with her streamlined nose high in the air, she exudes an air of haughtiness. Concorde knows she is the pride of the world's airliner fleet, as though she is a living entity. You can sense a strong, undaunted spirit in her as she quietly rests on the ramp awaiting the next dash across the North Atlantic at twice the speed of sound.

She is of good stock, and her ancestry dates back to 1783, when Joseph and Etienne Montgolfier invented the hot-air balloon. Their unmanned demonstration flights led to a command performance in Paris before the king and queen, a royal occasion for which the brothers hung a wicker basket below the balloon to accommodate a rooster, duck and sheep. The ascent was a great success; an observer, Jean-Francois Pilatre de Rozier, recognized the significance of it and volunteered to venture aloft. On October 15, 1783, before a huge and thrilled crowd in Paris, the Frenchman became the first human being to take to the sky. After centuries of effort, man finally was airborne.

Another Frenchman, Professor Charles, improved on the Montgolfier design and built a balloon that was inflated with the hydrogen gas recently discovered by the British scientist, Henry Cavendish. On December 1, 1783, Charles and Marie-Noel Robert performed a successful flight, whereupon Charles ascended by himself and rose to the incredible altitude of 9,000 feet.

On January 17, 1785, Frenchman Jean-Pierre Blanchard and American John Jeffries first crossed the English Channel by air when they flew a balloon from Dover to the coast of France. An attempt by the famed de Rozier to make a voyage in the opposite direction ended in tragedy when his balloon collapsed; the first man to fly became the first man to die in aviation.

In 1852, Henri Giffard of France significantly contributed to progress when he flew a steam-powered airship at the audacious speed of 6 miles per hour from Paris to Trappes. His design had been inspired by a model built by Pierre Jullien in 1850.

In 1857, Felix du Temple de la Croix patented a design for a steam-powered airplane. After the successful flight of a model replica, he built a full-size version, which briefly attained flight after running down a ramp to gather speed. Although there wasn't a practical control system and takeoff wasn't achieved under its own power, this marked the first time a powered airplane with a man aboard had left the ground.

Progress continued slowly but steadily; in 1898, Alberto Santos-Dumont rigged an internal-combustion engine to an airship and flew around the Eiffel Tower in Paris. The stage was set for somebody to build and fly the more practical airplane with a similar light-weight gasoline engine.

On December 17, 1903, Americans Wilbur and Orville Wright achieved the first true controlled flight in an airplane after painstaking and copious scientific experiments. Using a fragile biplane powered by their homemade, 12-horsepower, gasoline engine, the Wrights' first flight lasted 12 seconds and covered 120 feet. Three more attempts were made, and the longest flight managed 59 seconds in the air while covering 852 feet.

By 1905, the Wright Brothers were able to remain aloft for up to 38 minutes at a time in an improved aircraft. Unfortunately, it seemed that nobody in the United States was interested. The Wright Brothers put away their airplane and didn't fly again for almost three years.

Meanwhile, France had become the world's center of aviation. On October 23, 1906, Alberto Santos-Dumont flew his boxy biplane a distance of 200 feet; on November 12 he reached 722 feet at a speed of 26 miles per hour. The irrepressible Frenchman then built a small, but spirited, monoplane that he called the *Demoiselle*; he streaked along at 60 miles per hour—a phenomenal mile a minute.

In 1908, France welcomed the Wright Brothers to demonstrate their aerial prowess, which had become known to the Europeans. Although Orville remained at home to fly for the U.S. War Department, which finally had expressed interest in flying machines after being berated by Teddy Roosevelt for ignoring the matter, Wilbur went abroad and enthralled the air-minded French with consummate performances. By now, enthusiasm for flight in France had reached a fever pitch; in 1909, a quarter of a million people showed up at Rheims to watch no less than 35 contestants in the first air race ever held.

On July 25, 1909, Louis Bleriot made the first airplane crossing of the English Channel when he flew his Type XI monoplane from Calais to Dover. The Frenchman's design deviated from the Wright Brothers' concept of a forward-mounted tail and implemented the configuration that is standard to this day: The engine went forward and the tail went aft while the pilot sat in the fuselage with his hands on a control stick and throttle and his feet on rudder bars.

The year 1909 in France brought another milestone achievement in aviation—development of a lightweight gasoline engine that produced tremendous power for its size. Invented by a French engineer named Seguin, the Gnome powerplant was unique in that the cylinders rotated with the propeller while the crankshaft remained bolted to the airplane. This rotary engine was the world's first air-cooled engine; it eliminated the need for a radiator and cooling liquid.

Another "first" took place in France when Henri Fabre successfully flew a seaplane. Although history credits Glenn Curtiss, an American, with the honor of flying the first seaplane in 1911, the Frenchman took off from water at Martigeus in a Gnome-powered craft on March 28, 1910.

On September 29, 1913, the 200-kilometer barrier was broken when a French monoplane, the Deperdussin, established a speed record of 127 miles per hour with Maurice Prevost at the controls. This streamlined and sophisticated craft easily was 15 years ahead of its time.

No question about it—France had become the leader in aviation. But the colorful era of the pioneers ended with the outbreak of World War I in 1914; now, the development of the airplane took on a sinister nature—to destroy.

Britain and Germany went to great lengths to hasten their aircraft design and production. Some very capable machines rolled off assembly lines, including Sopwiths and Fokkers. France met the challenge, and its Nieuport fighters served with distinction. Its later Spad 13, built by the Societe pour Aviation et ses Derives, could fly at 130 miles per hour and up to 22,000 feet to take on the best adversaries from Germany. The famed American ace, Captain Eddie Rickenbacker, scored 14 of his 26 aerial victories in this craft, which he praised as "the ultimate aircraft of the war."

The United States lagged far behind in airplane development during World War I, charging into the conflict with bravado and bombast about thousands of American planes blackening the skies over Europe.

With the war over in 1918, aviation emphasis in Europe shifted to the inauguration of airline service. Transport aircraft built by French manufacturers such as Farman, Latecoeres and Potez—landplanes and seaplanes—complemented British-built de Havilland D.H.34s and D.H.66 Hercules airliners to pioneer the continent's network of air routes in the 1920s. Route extensions also were made across the Mediterranean and in South America.

Airline service in the United States languished until Charles Lindbergh and his *Spirit of St. Louis* performed their epic nonstop flight from New York to Paris in 1927. Overnight, Americans became fiercely air-minded; airlines launched passenger service throughout the land.

Soon afterward, an event took place in the United States that revolutionized airline service the world over: Donald Douglas developed the all-metal, twin-engine DC-1. On February 19, 1934, Jack Frye of Trans World Airlines and Eddie Rickenbacker of Eastern Air Lines took off from Los Angeles in the new transport and landed at Newark 13 hours and two minutes later—a new record for the transcontinental voyage.

An improved version was built for the airlines—the DC-2. Then came the legendary DC-3; airline executives clamored for orders, which ultimately would run into the thousands. The sturdy, reliable, high-performing Douglas airliners quickly dominated the world's air routes.

The air-transport industry in Europe, which had clung to traditional wood and fabric construction methods, took a back seat to the advanced American aircraft. Britain's Handley-Page *Hannibal*, a huge biplane with a cruising speed of 100 miles per hour, could not hope to compete. The French manufacturer, Marcel Bloch, tried to bounce back with the develop-

The United States captured supremacy of the skies when Donald Douglas developed the all-metal DC-1 in 1933. The DC-2 and DC-3 followed in 1934 and 1936 respectively. Seen here is the sole example of the DC-1 at Glendale (Los Angeles).
TWA PHOTO

Douglas DC-3 entered service in 1936 and quickly dominated air routes around the world. Operated by 90 percent of the U.S. carriers and 30 foreign airlines, the innovative transport carried 21-28 passengers at a cruising speed of 180 mph. Shown here is United Air Lines' NC16072 *State of California*.
UNITED AIRLINES PHOTO

Developed during World War II, the Douglas DC-4 was already a proven long-range transport when it entered airline service in late 1945. It could fly 42-60 passengers at 200 mph. Not pressurized, ceiling was limited to 10,000 feet. Illustrated is Pan American's N88951 *Clipper Racer*.
PAN AMERICAN WORLD AIRWAYS PHOTO

ment of his all-metal 220 and 221 twin-engine transports. Dewoitine produced all-metal 332, 333 and 338 models to try to protect the European and South American markets. But war clouds gathered over Europe, threatening not only the industry but nations as well. Hopes for peace ended when Germany invaded Poland in 1939; the following year, France was overrun by Wehrmacht troops.

The French aircraft industry was virtually destroyed by World War II. When the conflict ended in 1945, there was almost nothing left to build on. Yet, within a year, the Societe Nationale de Construction Aeronautique du Sud-Est in Toulouse rejuvenated a Bloch 161 design and produced 40 of these transports for Air France, which diligently pursued recapturing its prewar status as one of the world's leading airlines.

Britain also was behind in the field of transport development as a result of the war, having diverted all aircraft design to fighters and bombers necessary for the conflict. But a vigorous reconstruction program was launched to produce an airliner that would propel the British to the forefront of the industry.

Meanwhile, the United States not only retained its position as the world's leading airliner producer but achieved a virtual industry monopoly. Its four-engine DC-4 and Constellation, route-proven Douglas and Lockheed designs capable of crossing the Atlantic at 200 and 270 miles per hour, respectively, simply had no competition anywhere. The triple-tailed "Connie," in particular, with its unique aerodynamically tapered fuselage, had made a spectacular debut when Howard Hughes and Jack Frye of TWA flew the craft nonstop from Burbank to Washington, D.C., in six hours, 53 minutes.

Foreign as well as American air carriers rushed to equip their fleets with Douglas and Lockheed aircraft. The small series of stop-gap transports produced by the French immediately after 1945—however much they represented an admirable revival of an industry shattered by war—were non-contenders in a market dominated by the United States.

Assiduous research and development continued in Britain—something definitely big was brewing. Finally, the curtain was drawn to reveal an incredible achievement. On May 2, 1952, the British launched their de Havilland Comet—the world's first jet transport—into scheduled service with British Overseas Airways Corporation (BOAC). The performance of piston-engine, propeller-driven airliners was rendered obsolete by the epoch-making Comet; overnight, American domination of the market was in serious jeopardy. This was clear when Pan American World Airways ordered the British jets for its transatlantic routes.

Britain's gallant bid to wrestle air supremacy from the United States was short-lived, however, since several tragic accidents caused the Comet to be grounded for investigation. Metal fatigue due to stress from repeated pressurization cycles was determined to have caused the cabin to burst in flight, a problem that hadn't previously been experienced because the Comet's 30,000-foot

The pressurized Lockheed *Constellation*, also developed during the war, was launched into transatlantic service in January 1946. The first version carried 51 passengers at 270 mph, flew from New York to Paris in 20 hours, with refueling stops at Gander and Shannon. Seen here is TWA's 749A, N6014C *Star of Delaware.*
TWA PHOTO

On May 2, 1952 Britain launched its de Havilland *Comet 1*, the world's first jet airliner, into service with BOAC on the London-Johannesburg route. Note Boeing *Stratocruiser* in background—this American-built, piston-engine aircraft was utilized by BOAC for flights across the North Atlantic. A long-range *Comet* with transatlantic capability was in development, however.
BAe PHOTO

October 4, 1958: the British inaugurated the first transatlantic jet service with the long-range *Comet 4*.
BRITISH AIRWAYS PHOTO

cruising altitude had gone beyond the threshold of technical knowledge.

Although a program to ensure structural integrity was finished in time to allow Britain the honor of introducing transatlantic jet service with the improved Comet 4 on October 4, 1958, the position of leadership was recaptured by the United States only three weeks later. On October 26, Pan American inaugurated jet service from New York to Europe with the Boeing 707, a larger transport capable of carrying almost twice as many passengers as the Comet. The 707 also was 40 miles per hour faster with its 575 miles per hour cruising speed.

Soon after the Boeing 707 arrived on the scene, Douglas' rivaling DC-8 made its debut and was launched into transatlantic service by KLM on March 16, 1960. The world's airlines rushed to equip their fleets with the American-made jets, a cruel blow to the British Comet, harbinger of the Jet Age.

October 26, 1958: the United States recaptured leadership of the air, as Pan American began overseas flights with *Clipper America*, a Boeing 707-121.

TWA commenced jet service with the longer range Boeing 707-331 on November 23, 1959, from New York to London and Frankfurt. Seen here is 761TW, a B model with Pratt & Whitney turbofans. TWA PHOTO

The Douglas DC-8, America's second entry in the jet transport age, was launched into transatlantic service by KLM on March 16, 1960. KLM PHOTO

During this time, France hadn't taken a back seat in aeronautical development. Determined to regain the leading role it had in the colorful pioneering era, the French decided to introduce a new and radical design—one as revolutionary as the DC-3. This transport would be for the medium-range market, which like the long-range one, was dominated by the United States.

The eminent designer, Pierre Satre, prepared a bold draft of a jet airliner with three engines grouped in the tail—years before Boeing launched its similar 727 project. The French government approved the overall concept, but requested use of two engines when better propulsion became available.

Satre plunged ahead with elan. He designed the fuselage to be mated with an aerodynamically clean wing. He retained the rear-engine layout, which reduced fire hazard and provided better handling for pilots in engine-out conditions. In addition, the tail-mounted engines allowed passengers to enjoy an extremely quiet flight. To speed development and cut costs, he contacted the British and arranged to use the entire nose and cockpit layout of the Comet. This exchange between France and Britain foreshadowed future collaboration on a project of much greater magnitude.

This beautiful airliner was introduced to the world as the Caravelle. She performed her maiden flight on May 27, 1955, and underwent an extensive testing period before being granted a certificate of airworthiness. Air France also conducted route-proving trials to test the craft's integrity; the trials included four flights between Paris and Casablanca on one engine.

The Caravelle was launched into scheduled service on April 26, 1959. The world was astounded at her success, and orders poured in from numerous foreign carriers with medium-range routes to exploit. Then, an event occurred that many had considered impossible: United Air Lines purchased 20 of the unique jets for its domestic routes. At long last, France had penetrated the highly competitive U.S. market—and with no less than one of America's foremost airlines.

United's Caravelles entered service on January 18, 1961, and were immediately popular with the traveling public. Cruising quietly at 488 miles per hour, the comfortable and safe airliners linked 19 cities and carried almost a million passengers in the first year of service. Business boomed.

United's management was pleased with the reliability of the French transport and praised it: "The Caravelle is a sound, dependable aircraft with a high degree of mechanical integrity . . . we encountered no serious mechanical or technical problems, not even during the fleet introduction, which is usually the most critical phase of any equipment program."

Pierre Satre and all of France had reason to smile now—they were back in the mainstream of aeronautical development. But they didn't sit back and revel for

The *Caravelle* entered airline service on April 16, 1959, in Europe and was an immediate success. Subsequent stretched variants increased seating from 70 to 128 passengers. Although the French transport pioneered short and medium ranges in the jet age, production ended after 280 copies, while later American entries in the same market were produced in numbers beyond 4,700. AIR FRANCE PHOTO

Acquisition of a fleet of Caravelles by United Air Lines, largest carrier in the United States, was a major victory for the French air transport industry. Launched into service between New York and Chicago on January 18, 1961, the aesthetic aircraft performed with a perfect safety record before being withdrawn on October 25, 1970. Illustrated is N2001U (F-WJAP), actually the prototype Series 6R, used for demonstration flights in '61. UNITED AIRLINES PHOTO

long over their restoration of national prestige. However sweet the Caravelle triumph had been, it would have been sweeter had not Boeing and Douglas countered so effectively with their 720, 727, 737 and DC-9 short/medium-haul jet transports. Now, Satre and France embarked on a project that would unequivocally place the Gauls on top of the aviation world—the Super Caravelle, a supersonic transport.

Across the channel, the British were working on their project to recapture supremacy of the air. It was obvious that with the United States possessing 80 percent of the globe's airliner market, only a prodigious technical feat would haul the Union Jack above the Stars and Stripes. A supersonic transport was the answer.

A program had been set in motion to assess research and design studies on the SST. The program was based

Boeing 727, developed to operate in the short/medium-haul market, entered service on February 1, 1964. No less than 1,832 examples were built before production ended in August 1984. TWA PHOTO

Boeing ventured further into the short-haul market by launching the 737. Lufthansa was the sponsoring customer; United was first operator in the USA—April 28, 1968. Illustrated is N9004U in the airline's current paint scheme. UNITED AIRLINES PHOTO

Douglas developed the DC-9 for short-haul routes and was deluged with orders. Introduced in 1965, the twin-engine transport went on to become one of the most popular commercial jets in history. Seen here is TWA's N1051T, a DC-9-10, the first model to enter service. TWA PHOTO

BAC's Super VC-10 entered transatlantic service with BOAC on April 1, 1965. Powered by four Rolls-Royce Conway engines grouped together at the rear of the fuselage, the distinctive liner could carry 174 passengers at 560 mph. No airlines in the United States ordered it, however. BRITISH AIRWAYS PHOTO

at the Royal Aircraft Establishment at Farnborough. The road to laying groundwork for the development of a faster-than-sound airliner was strewn with obstacles—the major one being selection of a proper wing.

One design of a straight-wing aircraft the size of a 707 showed that only 15 passengers could be carried from London to New York, since the weight of fuel for sustained supersonic flight was horrendous. It became clear that a straight wing simply could not achieve the task, while a swept wing—as used on fighter planes—would require an incredibly high takeoff and landing speed out of bounds with reality.

The stymied British returned to their drawing boards to wrestle with the problem. Could the German scientists working at Farnborough discover a solution?

Dr. Dietrich Küchemann had worked with Professor Willy Messerschmitt on advanced high-speed projects in World War II. When the Russian juggernaut invaded Germany, Dr. Küchemann fled safely to the West. While von Braun and his Teutonic rocketeers went to America and abetted the intelligence for the space program, Küchemann and his clan of fluid dynamacists settled in Britain to conduct high-speed aircraft studies.

The Germans advanced the theory that a slender delta (a wing with a long chord and short span) would be the best shape for flight in both the supersonic and low-speed regimes.

British engineers were skeptical because initial American research had indicated that a slender delta would be extremely unstable in exaggerated flight attitudes, such as in taking off and landing. The Germans' advice was doubted.

Then Dr. Küchemann and a colleague, Eric Maskell, further unraveled the mystery when they presented an in-depth paper on the advantages of a slender delta in supersonic and low-speed flight. Their supposition was that boundary layer separation could stimulate a separation pattern that would generate growing vortex layers with an increasing angle of attack.

In effect, this bold theory explained—while traditional laws could not—why a child's paper airplane manages to fly with its sharp leading edges. Thus was provided a definite foundation on which to build, but doubts still persisted in the minds of engineers. Who was right, the Americans or the Germans?

A small aircraft, a Handley-Page 115, was built with a slender delta to assess this theory in actual flight. Test results were heartening—there was no instability even at exaggerated attitudes up to 20 degrees of pitch. The researchers realized that the Americans had been wrong all along. A slender delta could, indeed, accommodate low speed as well as supersonic speed.

Further testing in wind tunnels revealed that better

Dr. Dietrich Küchemann and his fluid dynamacists at the Royal Aircraft Establishment advanced the theory which led to the selection of the SST's wing.
RAE PHOTO

Pierre Satre designed the *Caravelle* and was the technical director of the French team in the Concorde project.
ACADÉMIE NATIONALE DE L'AIR ET DE L'ESPACE PHOTO

performance was possible with the ogival-delta wing (a shape that varies in curvature, first concave and then convex) for the problem of balance under flight. Also, loading conditions could more easily be controlled. A fighter outfitted with such a wing was test flown; handling qualities proved excellent. It was now, indeed, known that a supersonic airliner was a feasible possibility.

The British knew that the French also were sedulously working on a supersonic transport. At the Sud Aviation plant in Toulouse, Pierre Satre was engrossed in his Super Caravelle, with the support of the Office National d'Etude et de Recherche Aerospatiale, the French equivalent of Farnborough.

The British knew that an aeronautical project of such magnitude would require a huge influx of funds—perhaps more than they could handle alone—and sought to explore the possibility of collaborating with their French counterparts. The first contact was made on the industry level.

Pierre Satre was receptive to the idea of jointly producing the SST, for he, too, had become concerned with costs. An Anglo-French project would definitely be an effective manner in which to pool money and manpower—if they could agree on a common design.

There were immediate differences of opinion. The French wanted a medium-range SST; the British advocated a long-range version with transatlantic capability. Discussions continued between the industrialists to arrive at a preliminary design acceptable to both parties and to establish organizational details for the equal distribution of responsibility in the aircraft's development.

On the government level, President Charles de Gaulle earnestly committed France to the project. Indignant that the United States had seized domination of the world's aviation industry, he called for a technological offensive to challenge "the American colonization of the skies"—lest Europe become a subordinate. An achievement such as producing the fastest airliner in the world would restore France and Europe's prestige to superpower status.

Prime Minister Harold Macmillan, while less vociferous than de Gaulle, supported the project to secure Britain's future in aviation. But there was also an underlying reason—to gain British acceptance into the European Economic Community. Anglo-French collaboration on an SST would ease Britain's entry into the Common Market, whose membership was controlled by de Gaulle.

To coerce their sparring industrialists into finding a common design and setting a course of action—the discussions had stalled because of a dichotomy of opinion—

French and British governmental representatives exerted pressure. That resulted in bringing the long-range and medium-range concepts closer together, but talks about a final design dragged on.

Meanwhile, some observers in the United States who had laughed at the initial Anglo-French SST discussions—how could two nations who hadn't seen eye to eye on anything even consider sharing the development of such a complex aircraft?—now saw a potential threat when Paris and London urged action. Suddenly, it was learned just how much research already had been accomplished. With the Comet and Caravelle experience behind it, there was the realization that America's air supremacy could very well be threatened. Something had to be done immediately to stop this Anglo-French insolence.

A diplomatic offensive was launched from Washington, D.C. The first emissary, Eugene Black, former president of the World Bank, was sent to Britain. He met with the British minister of aviation and tried to talk him out of the SST project. This chat was followed by a steady stream of American inquiries questioning the SST's development.

The British were flabbergasted. They realized now that the pompous and patriotic de Gaulle had been right after all—the United States considered aircraft development across the entire world its exclusive preserve.

Back at the bargaining table, a final design on the size and shape of the SST had yet to be agreed on. Rounds of negotiations continued as frustration mounted in proportion to the fruitlessness. Tried of all the meetings, which now were crowded with bureaucrats, Lucien Servanty, Pierre Satre's senior designer, and Dr. William Strang, the British chief designer, fled from the restraints of officialdom and drove to the outskirts of Paris. They locked themselves in an office in the old Bleriot factory—the same building where French fighters were manufactured in World War I—and spent the remainder of the day resolving Anglo-French differences over the SST's specifics. At last, after numerous drawings, they agreed on a final design. It really was a rough draft, hardly more than a sketch, but it was enough to launch the project toward reality.

In view of the enormous funding that would be required, France and Britain decided the SST would be developed on a legal basis and drew up an international treaty that was registered at the United Nations. On November 29, 1962, the historic Anglo-French Supersonic Aircraft Agreement was signed in London. Only 20 pages long, it outlined the basis of collaboration "to develop and produce jointly a civil supersonic transport aircraft" and bonded the two nations together. There was no break clause—neither party could pull out. This highly unusual omission was Britain's idea to prevent France from possibly opting out at a later date.

On January 13, 1963, President de Gaulle held a press conference in the glittering surroundings of the Elysee Palace and announced to the world, ". . . Nothing would prevent the close relationship and direct cooperation, as these two countries have proved, by deciding to build together the supersonic aircraft Concorde . . ." This was the first time in public that the SST was christened with a name. In the same speech, however, he vetoed Britain's bid to enter the European Economic Community, making her future acceptance into the Common Market contingent on the SST's successful development. The chagrined British refused to spell "Concorde"—meaning a state of agreement—with the "e" at its end until 1967, when they would proclaim that it stood for excellence, England, Europe and entente cordiale.

Lucien Servanty, Pierre Satre's senior designer at Sud Aviation, worked out an agreeable design for the Anglo-French SST with the senior designer of BAC.
ACADÉMIE NATIONALE DE L'AIR ET DE L'ESPACE PHOTO

Dr. William J. Strang was BAC's senior designer.
BAe PHOTO

Sidebar to the *Comet* story: the British-built Vickers *Viscount* was the world's first turboprop airliner and entered service with British European Airways (BEA) on April 18, 1953, between London and Cyprus. A limited measure of success was then achieved in the United States. Eighty-two examples went into operation with Capital, Continental and Northeast Airlines on short/medium-haul routes. Seen here is N7408 after Capital (the largest *Viscount* operator) was absorbed by United in 1961.
UNITED AIRLINES PHOTO

Another British turbo-prop, the Bristol *Britannia,* began London-New York service with BOAC on December 19, 1957. No sales were made with American carriers. A freighter version, the CL-44 built by Canadair, later hauled cargo for Slick Airways, The Flying Tiger Line, Airlift International and Seaboard World Airlines. BAe PHOTO

The Lockheed *Electra* first entered service with Eastern Air Lines on January 12, 1959 between New York and Miami. Soon afterwards the aircraft was also operating with American, National, Braniff, Western, Northwest and other airlines. Although brought forth well after the *Viscount* and *Britannia*, sales of the *Electra* well surpassed those of the British-built propjets in the USA. Illustrated here is N6101A, the first of American Airlines' turbo-prop fleet.
AMERICAN AIRLINES PHOTO

Brian Trubshaw and John Cochrane, pilot and copilot of *Concorde* for Britain's first flight with the SST. BAe PHOTO

Chapter Two:
Development and Competition

So began the audacious assault on the sound barrier. The enterprise, mind-boggling in its vastness, began under the guidance of a committee of French and British governmental officials. They supervised the work of an airframe committee of directors and an engine committee of directors. Each was composed of an equal number of French and British engineering, production and sales experts, who, in turn, delegated more committees to cover support and the organization of almost a thousand subcontracting companies across France and Britain.

The main contractors were Sud Aviation, the British Aircraft Corporation (BAC), the Societe Nationale d'Etude et de Construction de Moteurs d'Aviation (SNECMA), and Bristol Siddeley Engines.* The French were responsible for the center of the aircraft, forward wing section, elevon control surfaces, landing gear, hydraulics, air conditioning generation, radio and navigation systems, flight-control system, aft engine components, exhaust nozzles, noise suppressors and thrust reversers. The British were responsible for the forward portion of the aircraft, the rear portion including the vertical stabilizer and rudder, electrical system, oxygen system, thermal and sound insulation, air distribution, fuel system, engine nacelles, controls, compressors, combustion chambers and turbines.

Although distribution of responsibility for the SST's development was made clear, trouble erupted soon after the start of the project. A common design had been agreed on, but the same dispute that hampered progress before the signing of the treaty now again strained the entente: The French wanted to build the craft for medium-range routes while the British called for nothing less than transatlantic capability. Initial plans were to build both versions, but now sufficient funding was available for only one.

Both parties were adamant in their stand and refused to relent. There was no chairman in charge to break the deadlock, so the matter had to be thrashed out by the committees. Language barrier aside, the situation was exacerbated by the different cultures of the two peoples; the Ecole Polytechnique-type discipline and protocol of the French contrasted sharply to the debating manner and phlegm of the British.

The British were not open-minded about anything but a long-range SST. They tried to convince their French partners that the essence of supersonic flight lay more in flying across the ocean in 3-1/2 hours than in flying from, say, Paris to Istanbul in 50 minutes.

Then, on June 4, 1963, Pan American World Airways placed options—provisionary orders—on six Concordes, to be used in transatlantic service. This was the first airline in the world to opt for the SST.

No less than the following day, President John F. Kennedy announced that the United States would build a superior supersonic transport. Said Kennedy: "If we can build the best operational airplane of this type, and I believe we can, then the Congress of this country should be prepared to invest the funds and the effort necessary to maintain the national lead in long-range aircraft—a lead which we have held since the end of the Second World War, a lead which we should make every responsible effort to maintain."

Then competition reverberated from the East; it was learned that development already had begun on a Russian SST. Veteran aircraft designer Andrei Tupolev and his son, Alexi, were steadfastly at work on a civilian delta-wing airliner in Voronezh (just southeast of Moscow).

A race was in progress! Now the French understood the British position and agreed that the long-range SST

---

*Sud Aviation became part of Aerospatiale in 1970; BAC became part of British Aerospace (BAe) in 1977; and Bristol Siddeley Engines became part of Rolls-Royce in 1966.

was the version to build. At long last, after 19 months of wrangling, the creation of Concorde began in earnest.

A giant industrial complex inexorably emerged across France and Britain—in Toulouse, in Filton, in LeHers, Farnborough, Melun-Villaroche, Bristol, Marignane, Patchway and in a hundred other cities. Tens of thousands of aircraft workers immersed themselves in setting up research, production and testing facilities to encompass hundreds of thousands of calculations, deductions, inductions and analyses for the refinement of a design that would be machined and assembled into supersonic airliners and then examined for integrity in the most comprehensive testing program in the history of aviation.

Soon progress was again impeded. In the face of the challenges from the West and East, French and British engineers decided it would be prudent to improve the performance of the long-range SST. Thus, they increased her length, wing area and gross weight to accommodate more passengers and fuel. The modifications and updated stressing for this improved Concorde, along with a major redesign of the engines to handle the additional weight, required a dramatic increase in the budget. This posed no problem for France, where de Gaulle reportedly regarded his financial officials as mere "quartermasters." But, in Britain, it was a different matter entirely.

Parliament rocked. When the revised estimate for developing the SST was revealed on July 9, 1964, members were aghast. The cost had doubled. When it became known that a "no-break" clause precluded abandonment of the project, SST opponents were incredulous and exasperated.

On October 16, 1964, Harold Wilson and his Labour Party gained power of the government and wasted no time in axing Concorde, which they considered an expensive and frivolous prestige project of the conservatives. An emissary was dispatched to Paris on October 28 to inform the French of London's cancellation of the SST. A brutal shock lay in store across the channel, however.

De Gaulle and France were incensed. Totally unsympathetic, they considered the maneuver a blatant disregard of devoir and produced the Anglo-French Supersonic Aircraft Agreement for a review of its terms. It was made clear that both nations were absolutely committed to the project. There was "no-break" clause, which was Britain's idea in the first place. Then, there being nothing more to say about the matter, the meeting concluded.

The emissary sullenly returned to London to report the French position. Now, Prime Minister Wilson was on the hot seat. In ensuing weeks, a stormy row broke out in Parliament between Concorde proponents and critics, while behind the scenes BAC executives called for the SST's survival for the sake of the aircraft industry and the thousands of people it employed.

On November 19, de Gaulle issued an ultimatum to the British government: Concorde *must* be built as specified in the treaty. To emphasize his firmness, he curtailed contact of the French aircraft workers with the British team, with future collaboration contingent on Britain's honoring her commitment.

While de Gaulle entrenched for battle, Wilson reviewed his options and found himself in a lamentable situation. If he continued with his original plan of action, a breach of the treaty undoubtedly would result in the matter being litigated at The Hague, where the International Court would most certainly rule in France's favor. The French then could continue with the SST's development, while Britain remained liable for half of the costs incurred—no matter what the total. The British also would have no right to utilize the craft on completion. And to cap this wretchedness, Britain's chances of entering the Common Market would forever be dashed.

On January 15, 1965, Wilson assembled the executives of the aircraft industry at his country residence for an uninhibited discussion of the ramifications of cancelling the SST. The picture presented was a somber one of BAC ceasing to exist and Britain's desecration as an aircraft producer. The prime minister weighed the opinions given and then acted: He scrapped the military HS-681 freighter, TSR-2 bomber and Harrier fighter projects, but capitulated to Concorde.

The giant industrial complex began to hum again as French and British workers plunged into their jobs with renewed confidence—and with a resolve born of the crisis to build Concorde as one team bonded together. The earlier language and cultural differences became immaterial now; the two peoples rallied in pursuit of the common goal to produce the fastest and finest airliner in the world.

Slowly, the machined components were transported across the lands to the assembly lines in Toulouse and Filton. By the spring of 1965, Concorde no longer was a paper airplane but an aluminum one. And her appearance promised performance.

The eminent designers, veterans of the Comet and Caravelle campaigns, were bold, but not extravagant, with their visions. Revolutionary as Concorde's design was, it also stayed within state-of-the-art technology, and thus, reality. Ideally, a swing-wing was more efficient, sweeping forward for low-speed takeoffs and landings and then back for supersonic flight, but its variable geometry was complex in nature—too complex, decided the Anglo-French team—to tackle at this time.

The stationary ogival-delta wing was thought a more practical solution—a compromise, yes—but it was elegant in simplicity and also precluded the need for the attachment of slats, spoilers and flaps. In addition, a Mach 2 speed limitation—1,350 miles per hour—was imposed. That eliminated the need for an outlandishly expensive titanium covering and increased cabin refrigeration to withstand the incredible temperatures created by kinetic heat of greater speeds. Concorde was covered with an aluminum alloy, with a temperature limit of 260 degrees Fahrenheit, as measured at the nose. There, kinetic heat could be this intense even with the outside air temperature as low as 75 degrees below zero. Someday, the redoubtable burdens imposed by Mach 3 and greater speeds would be overcome, but not now. Twice the speed of sound was fast enough.

As progress continued, so did spiraling costs. Another modification of fuselage and wing was undertaken to satisfy recommendations made by airlines for a transport with better passenger capacity. There also was the problem of developing the intricate Olympus 593 turbojet engines, an expense that was supposed to have been absorbed by the military before its similarly powered TSR-2 bomber was scrapped. And the aircraft designers again were calling on the engine people to produce more power for the additional weight.

Attention in Parliament anxiously returned to funding, which had become colossal compared to the original estimate. The latest bill was twice that of the last one—and there was no indication of what the ultimate cost would be. Critics advocated killing the project once and for all—regardless of the consequences—but they had to content themselves with hoping that some unforeseen technological barrier would repudiate supersonic travel. Prime Minister Wilson now was prepared to weather the crisis, since efforts were being renewed to enter the Common Market.

Across the channel, de Gaulle's "quartermasters" in the French Ministry of Finance were stupefied over the latest escalation. They looked for a scapegoat and chose the project's managing director, a Frenchman well liked by both British and French workers. De Gaulle acquiesced to the demands of his finance servants and ordered a change, but an outrage at Toulouse ensued, with employees staging a walkout to protest. The BAC people were equally distressed; the entire Anglo-French team had been dealt a heavy blow.

In the meantime, the Russians' progress on their SST continued. What's more, at the Paris Air Show, a model of the TU-144 in Aeroflot's color scheme was displayed. Although the Western press dubbed it the "Concordski" because of a similar appearance to Concorde, there were major differences between the aircraft. Like the Anglo-French team, Andrei and Alexi Tupolev stayed within state-of-the-art technology and produced an overall design that was not unlike Concorde's. But the wing was a double-delta shape, a large-scale rendition of the wing from the Saab Draken, a tried-and-true Mach 2 fighter developed by the capable Swedish. The engines also were grouped under the fuselage, as opposed to being separated and further out under the wing in the case of Concorde. Most distinctive about the TU-144 was its propulsion—four bypass turbofans, much less noisy than the turbojets of the Anglo-French craft. If the Russians could successfully build and operate a quiet SST, it would represent the pinnacle of air transport development.

Soon after the model display at Paris, the Russians invited selected British officials to see the TU-144 prototype under construction at the Tupolev plant. Western designers believed that stability problems would result from the engines being grouped under the fuselage. One of the Britishers related this to Alexi Tupolev. The young Russian concurred with the opinion, but went on to say that they couldn't control the engine-out yawing problem with the engines placed further out under the wing.

At the time, the engine-out condition at high speed also was a problem that the Anglo-French team was trying to surmount. This critical barrier to safe supersonic flight was well known to the aviation community as a result of the American B-58 experience. Built for the U.S. Air Force's Strategic Air Command, the four-engine Hustler set world records with its Mach 2 dash capability. On May 26, 1961, it flew nonstop from Fort Worth, Texas—with several in-flight refuelings—to Paris at an average speed just under 1,100 miles per hour. Eight days later, the same B-58 crashed with the loss of its crew. The Hustler was a formidable warplane when its engines ran properly, but there were numerous accidents. An engine failure at Mach 2 led to yawing in the dead engine, stress tearing off the tail, and destruction of the entire bomber. Flying beyond the sound barrier in multi-engine aircraft was precarious.

The specter of an airliner disintegrating at supersonic speed because of engine failure was daunting. The Tupolevs placed the engines together, at the expense of longitudinal stability, to prevent destruction from engine-out yawing. The Anglo-French team had yet to discover the solution for control with engines out under the wing.

Back in the United States, a competition between Boeing, Convair, Lockheed and North American was being held by the U.S. government to select an SST design for development. Preliminary evaluations based on 22,000 pages of data led to a standoff between Boe-

Tupolev Tu-144 was the first SST to fly, but was withdrawn from commercial service.
ATP/AIRLINERS AMERICA PHOTO

The Convair (now General Dynamics) B-58 *Hustler*, world's first supersonic strategic bomber, set 19 speed and altitude records with the U.S. Air Force. An engine failure at Mach 2, however, tore the aircraft apart.
GENERAL DYNAMICS PHOTO

ing and Lockheed, who battled each other down the homestretch to final evaluation and the construction of plywood facsimiles—mock-ups—of their supersonic concepts.

On September 29, 1966, Boeing revealed the mock-up of its SST with the fanfare of a Hollywood premiere. Airline executives, VIPs and reporters gathered in a dimly lit hall and waited for the show to begin. The lights came up and an unseen voice proclaimed, "This is the Boeing SST." There, for all to behold, was a white and yellow behemoth with a moving wing. The crowd applauded enthusiastically, then went aboard the mock-up. Saucy hostesses, to the accompaniment of background music, pointed out extrinsic niceties such as individual color television sets and "anthropometrically designed" seats. Supersonic flight was going to be plush, indeed.

Officially designated the Boeing 2707-200, this SST was awesome. At 318 feet in length, it was the longest airplane in the world. It would carry 292 passengers at 1,800 miles per hour, with thrust being provided by four powerful General Electric turbojet engines posi-tioned enigmatically under the tailplane. Titanium and stainless steel would be used for the covering to withstand the 500-degree temperatures associated with flight close to Mach 3. The radical and complex variable-geometry wing, swinging forward for low-speed takeoffs and landings and back for high-speed flight, was complemented with an array of 59 moving control surfaces—including triple-slotted flaps.

Lockheed's mock-up, a large version of the Anglo-French craft with a stationary double-delta wing, paled in homeliness compared to the Boeing SST. And both Concorde and the TU-144 now appeared as rudimentary as Santos-Dumont's fabric-covered Demoiselle.

The swing-wing craft fascinated the judges on the U.S. government's panel to select the final design. Boeing's exciting concept represented the leading edge in aeronautical technology—nothing else could match it. Of course, there was no question about it becoming a reality, for the company's impeccable track record was tangible testimony to its ability to produce innovative transports par excellence. Also, flying such a sophisticated airliner across the North Atlantic at 1,800 miles

Boeing's mock-up of its 2707-300 design revealed a conservative fixed wing. The earlier "Dash 200" with a swing-wing proved too complex to become a reality. DEPT. OF TRANSPORTATION PHOTO

Mock-up of Lockheed's bid for the U.S. SST competition, a 218-passenger craft with a stationary double-delta wing. LOCKHEED PHOTO

per hour with almost 300 passengers would be the perfect way to rebuff the arrogant Europeans. The world's airlines wouldn't be interested in purchasing their measly Mach 2, 100-passenger SSTs—they already were antediluvian.

On December 31, 1966, the final selection of the design to be developed as America's SST was announced: Boeing won. After a victory celebration, the company set in motion the machinery to transform the paper plane into a titanium and stainless steel one. Ominously, more money already had been invested in the United States SST program than into Concorde.

In Europe, the French and British engineers sighed with relief. They had been worried about the American competition being won by a plausible proposition, such as a larger version of Concorde, which they felt could have been a strong contender in the supersonic race. But Boeing's swing-wing concept was regarded as a brash and "bloody stupid idea," standing no chance of viability. It also was a clear symptom, they felt, of a disease that had been contracted through long-standing domination of the skies—overconfidence.

Progress was continuing on Concorde at a steady pace relatively free from political interference. Her basic airframe having been completed, attention was devoted to internal vitals.

The flight-control requirement was fulfilled by the unique "fly-by-wire" system—the first ever employed in a civilian transport and used elsewhere only in space capsules and advanced fighters. Herein, the pilot's movements of the control wheel and rudder pedals sent electrical impulses to hydraulic-powered actuators, which deflected correspondingly the six elevons on the trailing edge of the wing and the rudder aft of the vertical stabilizer. To provide realistic control pressure and to accommodate differing aerodynamics with changes in speed and its effect on stability, pre-programmed air-data computers integrated artificial feel and auto-stabilization into the system to enhance handling and stability. Trim adjustments were made to be effected through repositioning the data of the artificial feel system instead of deflecting conventional and speed-retarding tabs along the craft's streamlined surface.

Safety was assured through redundancy and informativeness. The electrical system was augmented with a separate secondary system—90 miles of wiring throughout the craft—to take over instantly should the first one fail. If both failed, a standby mechanical system of conventional cables and push-rods would allow the pilot to fly manually. A flight director, altitude-alerting system and a ground-proximity-warning system were installed with the standard flight instruments. Novel for civilian aircraft was the addition of an angle-of-attack indicator and a high-angle-of-attack warning system, which instantly indicated the margin above stalling speed, regardless of altitude and weight. This would provide for precision maneuvering and the recognition of high-pitch angles and sink-rate changes—the latter situation indicative of a wind shear condition.

An automatic pilot linked the flight system to the navigation system, which was composed of an ADF, a marker beacon receiver, two radio altimeters, two transponders, two magnetic heading reference displays, two weather radar systems, two VOR/ILS/DME systems for short-range navigation, and three computerized inertial navigation systems for long-range navigation.

Because of aerodynamics differing with speed changes—the center of lift moved rearward when accelerating from subsonic to supersonic, causing the aircraft to nose down—an effective method of balancing had to be devised. On a subsonic aircraft, nose-down pressure could be alleviated by deflecting an external trim tab. But such a protrusion on the SST would create prohibitive drag and an attendant reduction in speed and range. Although Concorde had a trim control to effect minor refinements in flight through repositioning data in the artificial feel system, the latitude of the shifting center of lift was such that greater corrective measures had to be taken.

Two methods to overcome the balance barrier were used. The first introduced camber and twist into the wing, which caused the center of lift to move forward at high speeds. This alone was not enough, so an ingenious fuel-transferring system was built to pump fuel aft to a trimming tank mounted in the rear of the fuselage. Changing the aircraft's center of gravity commensurately with the center of lift was a delicate task that had to be accomplished with care lest an out-of-balance condition occur. To assure control at all times, a frontal trimming tank was fitted ahead of the main tanks in the wing for a quick shifting of the center of gravity to a forward position, as would be required in an emergency deceleration. Powerful fuel pumps were installed to transfer fuel promptly at the rate of 2,000 pounds per minute. The flight engineer would supervise the transfer of fuel from his busy station behind the pilot and copilot; he could control the operation automatically or manually. For careful monitoring, the engineer and pilots' instrument panels were equipped with a Mach Meter and a center-of-gravity-position indicator to show the craft's balancing point and the permissible range for the speed being flown.

To achieve transatlantic range, Concorde's fuel capacity would account for half her total takeoff weight; her wing was filled with 13 tanks. When flying at supersonic speeds, fuel would be transferred from the outboard tanks to the inboard ones so the wing tips could flex upwards for better efficiency. Fuel would also serve another purpose besides feeding the engines: Its cooling properties would absorb the searing heat generated by Mach 2 speed. To test the reliability of the fuel system, a 350-ton facsimile was built and operated under scrutiny while subjected to different attitudes simulating flight.

Providing Concorde pilots with vision proved to be a major operation thwarted with obstacles. Her needle nose tapered to a point and was aerodynamically clean; a windshield was blended in smoothly with no indentation like that of a subsonic airliner. But technology could not provide a windshield capable of withstanding the burning heat generated by Mach 2 speed. Consequently, a metal visor was placed over the exposed glass. The visor would be lowered only for takeoff and landing, since it was assumed that the pilots wouldn't need to look out at other times. This cockpit turned out to be not only blind, but claustrophobic, so a tiny window was inserted to make amends. Now it was realized that the steep takeoff and landing attitude of a delta-wing craft would hamper sight of the runway, so another modification was deemed necessary. The streamlined nose was altered in shape to slant downward for a better view below, but it was not good enough, either. An innovative measure was attempted next—building a "droop nose" that would lower for takeoff and landing, with a two-piece visor fitted with small upper windows and internal periscopes. It was at this point that the U.S. Federal Aviation Administration intervened and bluntly avowed that the craft would not be allowed into the United States for safety reasons if better visibility wasn't provided. The FAA's adamant stand, heeded by the designers and applauded by pilots who were to fly the SST, provided the impetus for advancing glass technology. A glazed, heat-resistant visor was quickly developed and installed over the "droop nose." Concorde could finally "see."

The painful lesson of the Comet—pioneering the Jet Age with tragic accidents marring its breakthrough of the unknown—called for caution in assailing the greater threshold of supersonic flight. The aerodynamics of the subsonic regime were understood down to the last 0.01 percent, but the territory beyond the sound barrier was relatively unknown and unexplored. True, advanced military planes with pilots in pressure suits could fly at Mach 2, but their hell-bent dashes were confined to short duration. An airliner filled with mufti-clad passengers rocketing through the stratosphere faster than most fighters on vast aerial journeys across the oceans was a more formidable feat—a mind-boggling one.

Obsessed with safety, the Anglo-French engineering team spared no effort to reduce the risk of error. Ultimately, almost a third of the entire $3 billion project was devoted to assess Concorde's integrity for ensuring utter reliability. Precaution was paramount.

Critical questions had to be answered. Would the strain associated with repeated cycles of cabin pressurization cause metal fatigue, weaken the structure, lead to a structural failure and an in-flight explosion? Would the rest of the airframe's components bear up under the stress of supersonic flight? Also, since it was known that high temperatures could weaken aircraft skin materials, what effect would the searing heat of Mach 2 flight have on Concorde's covering? Furthermore, after expanding under extended exposure to intense heat, would the covering return to its original shape or deform when the craft returned to the lower temperatures of subsonic flight? How would the repeated interaction of heating and cooling affect Concorde when she also was subjected to the externally applied mechanical loads inherent with actual flight?

A full-scale static test was conducted at Toulouse. There, infrared heaters and liquid nitrogen exposed the airframe to alternating cycles of heating and cooling, while hydraulic jacks applied mechanical loads simulative of those imposed when the craft was taking off, climbing, cruising, encountering turbulence, descending and landing. Her dynamic reactions recorded, it was determined that Concorde's aluminum alloy covering—Hiduminium RR58, a composite of aluminum, copper, magnesium, iron, nickel and silicon—could withstand the thermal exposure and retain its shape after cooling. A mechanical weakness was revealed in the vertical stabilizer, whereupon the structure was reinforced.

Concorde's immediate future was filled with promise, but how would years of service affect her well-being? To find out, the Royal Aircraft Establishment at Farnborough began building a stupendous diagnostic rig to superimpose the craft and duplicate on her structure the exact sequence of thermal and mechanical stresses involved in a supersonic flight cycle. A computerized monitoring system would continuously record thousands of measurements for analysis. When completed, this elaborate mass of machinery, in itself a brilliant feat of engineering, would be the most comprehensive aircraft-testing rig in the history of aviation. The most scientific examination of Concorde's integrity would continue indefinitely, revealing that the craft could fly soundly into the 21st century. Safety had, indeed, become an obsession.

The heart of any airplane is its power plant. Concorde's four Olympus 593 engines were state-of-the-art turbojets capable of producing brute force. The options at the beginning had been either a pure jet or a quieter bypass turbofan. The latter was quickly ruled out. Larger and weighing 2,000 pounds more than a pure

jet—four engines would add an extra and unwanted 8,000 pounds on the weight-critical SST—a turbofan could only produce a thrust velocity of 1,500 feet per second. That wasn't nearly enough to propel a craft intended to fly at 2,200 feet per second. The Olympus turbojet, on the other hand, was smaller, lighter and had an impressive velocity of 3,000 feet per second, which was sufficient to hurl Concorde beyond the sound barrier.

Serious problems had been encountered, however, when designers had increased the size and gross weight of the craft. Extracting more power than what already was available had resulted in a frustrating and time-consuming redesign of the Olympus; 10 variants employing the latest technology were built before the final model provided the needed pounds of thrust. What helped immensely was the attachment of an afterburner, which injected raw fuel into the hot exhaust gases for instant combustion and tremendous power. Alas, use of afterburners on all four engines would quadruple fuel consumption to an exorbitant 3,000 pounds per minute—outrageous for continuous operation. With fuel critical for transatlantic range, the afterburners would be employed judiciously only for takeoff and when accelerating to supersonic speeds.

The most important components of the SST engines were their exhaust nozzles. Whereas a subsonic jet engine had one convergent nozzle through which burning gases are expelled, the Olympus had convergent and divergent (narrowing and widening) nozzles to produce extra power. To optimize thrust in every regime of a flight cycle, the exhaust nozzles were not fixed in a stationary position but made variable to accommodate changing conditions. It remained to be seen how well air speeds in the intake and the pressures and temperatures in the engine and nozzles would harmonize in conditions varying from standing still on the runway to supersonic flight at 60,000 feet.

The power-packed Olympus 593 also introduced a novel feature for aircraft engines—an electronically automated control system with a back-up manual system. The pilot would be able to select, with switches, the mode of power he chose—takeoff, climb or cruise. The proper thrust automatically would be provided. If necessary, he could revert to the conventional manner of controlling power by hand adjustments.

Thousands of hours were devoted to testing the engines in simulated Mach 2 flight conditions, with mechanical blowers forcing intensely heated air into the intakes. A potential problem of shaft breakage was detected and corrected. Titanium blades were installed in the compressors to withstand internal temperatures of more than 1,000 degrees. Special synthetic lubricating oils were developed to cool the components, which ran twice as hot as in a subsonic engine. Variable intake ramps also were devised to retard air flowing supersonically into the engine down to subsonic speeds for efficient consumption.

Although great strides had been made in the SST's development, progress had been hindered by the initial disagreement over the range of the craft, increasing her size and the ubiquitous technical problems associated with supersonic speed. Disappointment mounted as the maiden flight lagged increasingly behind schedule.

On December 11, 1967, a palliative ceremony took place at Blagnac Airport in Toulouse involving more than a thousand political dignitaries, technocrats and reporters. In a freezing wind, statesmen from France and Britain spoke of the importance of the day's event, which would provide visible proof of Europe's determination to regain rule of the skies. When they finished speaking, bands from both nations struck up each other's national anthem, "God Save the Queen" and "Marseillaise." Then, two large hangar doors slowly parted, revealing, in her first public appearance, Concorde.

Gleaming in white with red trim, she was sleek and beautiful. The crowd was caught up in the emotion of the moment. Here, after five years of agreements, arguments, collaboration and recrimination, was a supersonic transport for the entire world to see.

Immediately after the first roll-out of Concorde, dark clouds again gathered over the project. Britain, in dire economic throes, was facing financial disaster unless immediate and drastic cuts were enacted in domestic and foreign expenditures. The government's axe fell swiftly: All military forces outside Europe, save for Hong Kong, were recalled; new fighter planes scheduled for the Royal Air Force were cancelled and the "free" National Health Service would start charging for prescriptions.

Anti-SST members of Parliament were outraged that Concorde had been spared when other sacrifices had to be made. They criticized the continuance of a project still spiraling in costs. Also, there was the discovery that Concorde was almost 6,000 pounds overweight. She would be able to carry only 10 passengers from London to New York! As engineers scampered to study the mortal problem and find a solution, government officials threatened to abandon the project if the weight wasn't reduced.

A pall of doubt spread over Concorde. Adding to the uncertainty of her future was President de Gaulle's rejection of Prime Minister Wilson's attempt to secure Britain's membership in the European Economic Community. The supersonic entente was severely strained by this lack of trust.

France, meanwhile, notwithstanding the wailing of her finance ministers, remained stubbornly committed to Concorde. But French leaders realized that the SST's fate now rested solely on the actions of the British government. Following de Gaulle's second veto of Britain's bid to enter the Common Market—a stinging rebuff that had done nothing to foster continued esprit de corps between the two nations—the French hardly were in any position to exhort their counterparts across the Channel.

Dissatisfaction with the hoary, patriarchal Gaullist regime was stirring across France. Perhaps the old general's anglophobia and grandiloquence about returning France to relive "la gloire" of Louis XIV's era was becoming regarded as purely anachronistic. Public discontent surged when students of the Sorbonne went on a rampage through the streets of Paris while disillusioned aircraft workers walked off their jobs.

As discord pervaded France and Britain, Concorde sat desolately in her hangar, waiting for better days—if they were to come.

Back in the United States, Boeing's braggadocio about a superior swing-wing supersonic transport had turned to buffoonery. Fabricating tenacious titanium into vast, complex sheets, which was necessary to provide a covering capable of withstanding 500-degree temperatures caused by Mach 3 speeds, was a problem that hadn't been solved because production hadn't even begun. The greater problem of finding a workable solution for operation of the variable-geometry wing—given the plane's unusual configuration—had Boeing stumped. Owing to the aft positioning of the four engines under the tailplane, a serious mass-distribution problem was created. This was accompanied by instability in the low-speed regime. To counter the dilemma, a canard—a forward horizontal stabilizer—was mounted on the fuselage. This aerodynamic refinement ambiguously induced instability at all speeds, which, in turn, educed flexing and straining of the airframe, requiring structural reinforcement with a penalty of additional weight.

Weight so plagued the Boeing 2707-200 that it ballooned out of proportion with reality. The massive pivots and complex actuator mechanisms in the swing-wing weighed 40,000 pounds alone. Unforeseen structural reinforcements, including strengthening the wing itself to support the weight of the swinging apparatus, compounded viciously until the recondite design was 75,000 pounds overweight. With such a burden, the craft wouldn't be able to carry any passengers across the Atlantic.

Out of necessity, the wallowing wonder was subjected to an assiduous weight-reduction program. All superfluous items were removed, including television sets. Alas, the measures taken were inadequate: The design still was 52,000 pounds too heavy.

After 30,000 hours of research in eight different wind tunnels, 80,000 hours of computer analysis, and more than eight million hours of human engineering, the company conceded that the swing-wing supersonic transport—designed to become the largest, fastest and most luxurious airliner in existence—would not take off from the drawing board much less runways in New York, Paris and London. Hitherto the world's pre-eminent manufacturer of air transports, Boeing was stultified.

Regrouping quickly to launch another offensive—and to try to save face—the company called on its designers to present the feasible fixed-wing concept for its SST bid. The design produced, the Boeing 2707-300, was remarkably similar to the Lockheed version rejected in the competition.

The U.S. government, which was funding almost the entire project and which had set the criteria for the SST's performance, reluctantly concurred with developing the less prodigal craft to save America's contention in the supersonic race. Quietly, the mighty swing-wing SST was banished to oblivion.

Trying to efface the stunning technological defeat, Boeing overnight began voicing the virtues of the new fixed-wing supersonic transport. The public was con-fused by the sudden change; congressmen, worried over the horrendous expenditure, launched a major campaign to scrutinize what they now considered to be a delusive project. Estimates for the SST's development had spiraled from $1 billion to more than $4 billion. Casting another shadow over the American SST's future, vociferous environmentalists predicted that supersonic planes would harm the ecology and mounted an assault to stop the project. Joining ranks with the congressmen who were outraged by the ex-travagant funding, they plotted to kill the SST.

Meanwhile, on the other side of the world, Soviet engineers were preparing the TU-144 for flight. The pace of progress had continued swiftly and unabated since the Council of Ministers of the Supreme Soviet launched the Russian SST project in 1962. It was a prestige project with a practical application.

Air travel had played an important role in allowing a vast country with limited surface transportation to be traversed. So great was the need for a network of air routes to link great distances that Aeroflot had been developed into the largest airline in the world. Super-sonic flight would allow linking even the most remote cities.

On December 31, 1968, at an airstrip outside Moscow, the Tupolev father-and-son team was joined by bureaucrats, technocrats, Aeroflot directors and Air Force officers to witness the fruits of six years of diligent labor. All eyes focused on the TU-144 sitting at the end of the runway, where test pilot Eduard Vagonovich Elyan and his crew went through the perfunctory checklist to assess the aircraft's systems. Then, power was added and the needle-like SST taxied and aligned with the center of the runway, indicating that the moment of truth was at hand. Full power surged in the four Kuznetsov NK-144 turbofan engines, emitting a muffled thundering across the landscape. The praying mantis shape moved slowly at first, but gained speed gradually and surely, its afterburners spewing tremen-dous thrust. Faster . . . faster . . . Now the speed was great and the SST was committed. The nose wheel lifted off the ground; the main gear followed; the TU-144 was airborne, climbing smartly for altitude, the first SST to take to the air.

It was a moment of jubilation for the elderly Tupolev, the crowning achievement in a long and productive career. His TU-104 had launched Russia into the Jet Age when Aeroflot had introduced the airliner in scheduled service between Moscow and Irkutsk on September 15, 1956. Now, he had led Russia into the Supersonic Age in the air-transport industry. But he would not see his SST go into service with Aeroflot; he died soon afterward, leaving the project's future in the hands of his son.

Back in Britain, Concorde had survived once again. Her engineers had eliminated the excess weight by redesigning the aft portions of the engines with the aid of an American company, the TRE Corporation of California. This valuable transatlantic exchange was enhanced by the contribution of Jack Andresen, a brilliant engineer who was working with the Intercon-tinental Dynamics Corporation of Engelwood, New Jersey. Andresen, holder of more than 30 patents, designed the SST's primary airspeed indicator, a specialized instrument because of the craft's 1,340 mile per hour speed and wide-ranging operating domain. It possessed the dual function of providing airspeed infor-mation and the maximum allowable airspeed for exist-ing conditions.

With the weight problem solved, the British govern-ment approved continuation of the SST project. Prime Minister Wilson was intent on using Concorde as the ticket of entry into the Common Market. France could hardly dare to slam the door in Britain's face a third time. Yet, there also was an underlying reason for sup-port of the SST: After all the military and social cut-backs forced by the economic crisis, cancellation of the most daring venture would shatter morale across the land. The Union Jack had to stand for something, and flying high on a supersonic transport it would represent Britain's technological superiority.

Development had commenced again in the autumn of 1968, when French aircraft workers had decided to go back to work. Havoc was wreaked on the original schedule, however, dashing Concorde's chances of being the first SST to fly. In fact, the maiden flight now was fully two years behind the target date established back in 1962.

The project back in motion, safety, as ever, was heeded with painstaking testing. The Anglo-French airworthiness authorities and manufacturing officials went abroad to work with America's Federal Aviation Administration (FAA) and the National Aeronautics and Space Administration (NASA) to investigate performance projections of SSTs and Concorde. At NASA's Ames Research Center at Moffett Field in Sunnyvale, California, Concorde's behavior in flight was analyzed on the Flight Simulator for Advanced Aircraft—a computer that dealt out every conceivable parameter.

Taking the investigation a step further, SST performance was simulated in actual flight with a variable-stability Mirage in France and with a Convair Total In-Flight Simulator in the United States. Curious landing characteristics were revealed: "Flying on the back side of the drag curve," pilots called it. In a subsonic airplane, if the pilot pulled back on the control wheel while maintaining constant power, the airplane would initially climb. In Concorde, if the pilot pulled back while maintaining constant power, the craft would descend; if he pushed forward, the craft would climb. This behavior was exactly the opposite of conventional practice, and it was readily understood that proper power application would be a significant factor in flying the SST.

Concorde was turned over to André Turcat, chief test pilot of Sud Aviation in France. Numerous sprints were made up and down the runways at Blagnac Airport in Toulouse to assess the integrity of the controls, engines and innovative, computerized, carbon brakes— the first ever on a civilian aircraft. For weeks, the SST underwent ground testing. The crowd that came out daily to observe wondered when the momentous takeoff finally would occur.

On March 2, 1969, Turcat felt the moment was at hand. Before a huge crowd of citizens and reporters, he taxied Concorde to the end of the runway. Next to him was co-pilot Jacques Guignard. At the flight engineer's station was Michael Retif, while the observer, Henri Perrier, took the jump seat.

Cool and methodical, Turcat was an experienced

André Turcat, chief test pilot for Sud Aviation and Aerospatiale, directed the French flight program.
ACADÉMIE NATIONALE DE L'AIR ET DE L'ESPACE PHOTO

veteran who was used to rising to the occasion. An ex-military pilot, he had, in 1954, become the first European to break the sound barrier. Numerous speed records followed, and in 1959, then Vice President Richard Nixon presented the Frenchman with the distinguished Harmon Trophy for his high-speed flight contributions. Now, the Chuck Yeager of Europe would be calling on all of his valuable experience for the greatest flight of all.

Concorde's four Olympus turbojets roared in unison as the plane turned onto the runway and was cleared for takeoff. Concorde accelerated slowly at first, but gained speed with each passing second, looking like a determined and huge pterodactyl in search of resurrection. The ground raced by . . . 1,000 feet . . . 2,000 feet . . . 3,000 feet . . . 4,000 feet. . . . Turcat pulled back on the control wheel, breathing life into the speeding craft. Her nose pitched sharply into the air and seconds later her main landing gear lifted off. Concorde climbed steeply into the sky, her engines ablaze with orange brightness and emitting a thunderous crackling that reverberated across the field to compete with the roar of thousands of cheering French who were enraptured by the birth.

Concorde skirted around Toulouse—where almost 200 years earlier the Montgolfier brothers had demonstrated their balloons for the citizens—and climbed toward open country. Turcat felt out her controls and responses and was pleased with her soundness. The baby was healthy. Limiting the maiden flight to short duration, he returned to Blagnac Airport, mindful of the lessons learned in the simulator about proper power control when flying on the back side of the drag curve. Concorde approached nose-high, typically delta style, touched down on the runway with grace, and rolled to a stop in 8,000 feet.

All of France was electrified. "Concorde" was the household word across the land, with Concorde cafes and a full-scale mock-up built next to the main terminal building at Orly Airport in Paris. Support of the SST project by the French was virtually total, and it evoked intense feelings of national pride in all walks of life.

In the United States, Pan American World Airways launched advertisements displaying Concorde in its color scheme, letting it be known which airliner they soon would be flying.

On April 9, 1969, it was Britain's turn to fly Concorde. Brian Trubshaw, an ex-Royal Air Force pilot with 25 years of test flying in his logbooks, lifted the SST off the runway at Filton and headed for the RAF base at Fairfield. While approaching to land, however, he discovered that the radio altimeter had malfunctioned. With the pilot sitting 40 feet in front of the nose wheels, landing a delta gracefully by feel and judgment alone was akin to stepping down a stairway in the dark. Trubshaw, the stalwart professional, babied Concorde in by the seat of his pants for a perfect landing.

Now it was Britain's opportunity to bask in the limelight, as news reports flashed across the land. BBC television programs were interrupted to show a replay of Concorde's takeoff. Public attention was spectacular. Pride swelled in the majority of citizens.

In Parliament, pro-SST members were elated by the successful performance. They knew that now the project had to continue in earnest. Critics, who had been hoping that some unforeseen technological barrier would stymie the craft, lamented that Concorde had just presented them with a *fait accompli*.

View of *Concorde's* flight deck shows pilot, copilot and flight engineer stations. Although "busier" than a subsonic cockpit, layout of instrumentation is comfortable for the crew. BAe PHOTO

Jack Andresen from the Intercontinental Dynamics Corporation of Englewood, New Jersey, designed the specialized airspeed indicator for Concorde. IDC PHOTO

Moment of truth: *Concorde* (prototype 001, F-WTSS) lifts off from the runway at Toulouse for her maiden flight on March 2, 1969. ANDRÉ TURCAT PHOTO

In-flight view of *Concorde* on her first flight. Landing gear was locked down for added safety, and speed was held down to 295 mph (supersonic dashes came later). ANDRÉ TURCAT PHOTO

Concorde performing her first landing—brake parachute was deployed to shorten roll-out. ANDRÉ TURCAT PHOTO

G-AXDN, the first pre-production *Concorde*, is illustrated on one of her many test runs. BAe PHOTO

Chapter Three:
Testing and Opposition

In June 1969, Concorde performed at the Paris Air Show. Then, she proceeded to undergo examination in the most exhaustive flight-test program in the history of aviation. The French were responsible for honing handling characteristics, while the British concentrated on extracting maximum performance. The threshold of the sound barrier commanded utmost caution in the exploration of unknowns and deciphering them into understandable phenomena. No less than a staggering 5,495 hours of flight-testing would be devoted to safety and obtaining the Certificate of Airworthiness from French and British aviation authorities—the prelude to carrying paying passengers.

Grooming Concorde for airline service led to 3,500 tedious modifications. The most antagonizing aspect of the program was discovering how to ensure smooth operation of the engines in all flight regimes—from takeoff to Mach 2 cruise at 60,000 feet.

One of the problems that became apparent was the differing of aerodynamics and airflow along the sweeping wing at the inboard and outboard engine positions. To compensate for this, each intake was angled in slightly and given a different geometry to absorb a uniform airflow for maximum engine efficiency.

Everything worked fine until passing through the speed of Mach 1.4, where a phenomenon known as "surge" came into effect. Surge is an explosive-sounding backward-spilled pressure wave caused by a breakdown of smooth airflow into the engine. To control airflow into the engine, which had to be decelerated from supersonic speeds to Mach 0.5, variable-geometry intake ramps were raised open for subsonic flight and lowered for supersonic flight. Surging occurred when the ramps were positioned incongruously with the existing angle of attack, speed, pressure and temperature. The problem required an exhaustive investigation into measuring the effects of attitudes, speeds, pressures and temperatures on airflow, then translating these measurements into control laws governing exactly how the ramps should be positioned in each regime of flight. A further modification later was made to increase the frontal area of the intake, which totally eliminated the occurrence of surging even if the aircraft was mishandled with pronounced attitude changes. The engines were assured of running perfectly at all times—an outstanding achievement in view of the drastic condition changes during a supersonic flight cycle.

Since the Mach number is a function of temperature, the automatic pilot system had to be modified. When flying through areas of temperature change, the pilots discovered that the automatic pilot would pitch the craft sharply up or down to pursue the Mach number it was programmed to maintain. The system was modified to hold the speed programmed into it—without excessive pitch changes, but by smoothly adjusting the throttles.

The most daunting threat to safe supersonic flight was eliminated early in the testing program. The dire yawing problem associated with losing an engine at supersonic speeds was eliminated by gearing the inner pair of elevons to have a movement of roll to a movement of pitch ratio of .07 to 1.0. Stability now was so superlative that if Concorde sustained a double-engine failure on one wing, she would merely roll lazily to the side and could easily be corrected by leveling the wing and nudging the rudder. Anglo-French aviation authorities, indurated from the B-58 accidents, were astonished at the SST's docility in such a condition that hitherto was totally destructive. One British official remarked that a boy of 16 could handle the emergency—qualitative testimony to Concorde's safeness.

While flight-testing was under way, an insidious threat from outside the government was manifesting

itself like an ineradicable gangrene. The Anti-Concorde Project, composed in large part of university liberal arts professors, attacked the SST as a technological folly that would destroy the earth's upper atmosphere and environment. Organized in Britain and partial to using newspaper advertisements to publicize their hostile views, the critics gathered support and mounted an all-out assault to kill Concorde.

In France, attacks on Concorde and predictions of environmental doom generally were met with stony indifference. The single-minded French, fiercely proud of the SST, looked on the Anti-Concorde Project as an insensate, unpatriotic radical group.

Back in the United States, environmentalists now were regarded as a highly organized and powerful conglomeration capable of exerting tremendous influence. Under the banner "Coalition Against the SST," environmental groups—including the Sierra Club, Friends of the Earth, Citizens League Against the Sonic Boom, the Wilderness Society and the Committee for Green Foothills—had united efforts to stop the American SST project and had launched a stupefying propaganda barrage.

The alleged consequences of supersonic flight were as diverse as they were numerous: The SST would break windows, crack walls, stampede cattle, "hasten the end of the American wilderness," destroy the ozone layer and inflict cancer on a defenseless humanity, damage unborn children, cause hearing loss and changes in the endocrine system, create disturbances in the cardiovascular and respiratory systems, adversely affect the weather, melt the ice caps and flood the Earth (or bring forth the next Ice Age—opinion varied), deplete the world's resources and disturb fish off Newfoundland. . . .

SST opponents urged citizens to "write, telephone and wire your congressmen" for the project's cancellation.

Meanwhile, all was not well at Boeing. Vexed engineers were finding their delta design as refractory as the discarded swing-wing concept—it was plagued with weight problems! This time, the cause was attributed to the specialized titanium covering. Boeing was becoming devitalized by the supersonic transport challenge; major contractors began pulling out of the project.

More and more members of Congress were aroused by the huge expenditure of public funds on an aircraft not yet off the drawing board—and one that was inciting such a volatile reaction from the environmental lobby. Political criticism of the SST project grew and intensified; its death appeared imminent.

President Richard Nixon, who had inherited the project and who was strongly in favor or it, shrewdly maneuvered to keep the SST alive. When additional

Prototypes flank pre-production Concorde. The original metal visor, as seen clearly on the craft towards the far right, restricted pilots' forward vision. BAe PHOTO

funding was required for the aircraft's continued development, an allocation for the SST was packaged as part of an appropriation for the Department of Transportation. Failure of Congress to approve it would have halted federal funding of highways and railroads as well.

The House of Representatives narrowly passed the appropriation, but the Senate voted to delete the allocation for the SST. An affronted President announced to the nation, "What is involved here is not just the 150,000 jobs that will be lost if we don't build it; but what is lost here is the fact that the United States of America, which has been first in the world of commercial aviation from the time of the Wright Brothers, decides not just to be second, but not even to show."

An attempt to save the SST was motioned in the House, but the Senate refused to consent and the American supersonic transport officially died on May 19, 1971. After eight years of effort and an expenditure totalling half of what the entire Concorde enterprise cost, the United States didn't have a single, tangible piece of airplane to show for it.

With the Boeing SST out of the race, it became unmistakably clear that America now was running the risk of relinquishing her long-held leadership in aviation to the Europeans. The TU-144 wasn't considered a mortal threat, however, for, notwithstanding the loss of face the U.S. government would experience if a Russian SST pulled up to the gate at Kennedy Airport, it was doubted that the craft could penetrate the market outside the Soviet bloc. But the Anglo-French SST was a definite menace, for she already had proven her excellent flying qualities and production soon would be going full bore. A governmental ad hoc committee discreetly probed expedients that could be taken to undermine France and Britain's challenge of America's air superiority.

The committee's report was shockingly malicious: ". . . U.S. noise standards could conceivably bar Concorde from access to the principal U.S. airports, which would undoubtedly doom the Concorde program . . ." Nixon was outraged when he read the report. He abhorred the idea of resorting to political sabotage to salvage America's prestige in the face of what he considered an admirable achievement on the part of two staunch allies. Diametrically opposed to his advisers, he embodied the American spirit of free competition and gave his personal word to the leaders of France and Britain that Concorde would not be subject to banishment because of xenophobia. "In the United States," he assured them, "you will get a fair shake."

Exaggerated view of Concorde's ogival-delta wing. Below, on wheels, innovative carbon brakes were the first ever used on a commercial air transport.
AIR FRANCE PHOTO

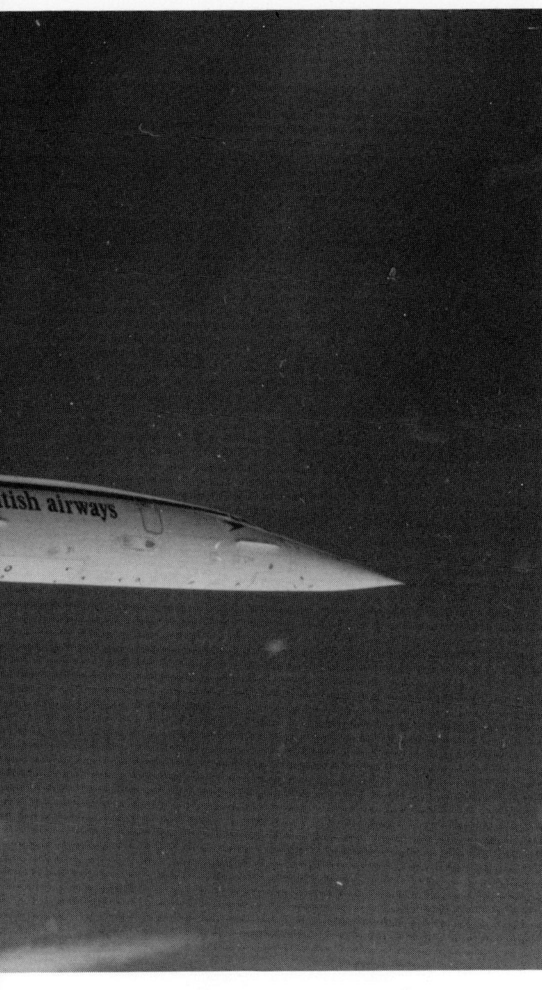

Top: Painstaking testing ensured smooth and efficient operation of Concorde's Olympus 593 turbojet engines from on the runway to twice the speed of sound at 60,000 feet. BAe PHOTO

Left: Concorde underwent 5,495 hours of flight testing, more than 3½ times as much as the 747, and more than any other aircraft in history. Note droop nose with glazed visor—FAA of the United States induced excellent visibility for the SST. BAe PHOTO

Across the ocean, progress had continued smoothly. In June 1970, Edward Heath became prime minister and the British Conservative government returned to power. Both were friends of Concorde and had no intention of cancelling Britain's commitment in her development. Manufacturers and SST proponents found their confidence soaring with this political transition; it was certain that nothing would stop the project now. Heath also immediately initiated negotiations for Britain's acceptance into the Common Market, and this time the French not only were friendly, but even encouraging!

Since Concorde had been blasted by the vociferous Anti-Concorde Project as a hideous demon that would wreak havoc on the ground with her sonic booms, the British government scheduled a series of supersonic flight tests to alleviate its citizens' fear of noise damage. Concorde was flown faster than sound along the west coast of the country. Alas, the test program backfired: Numerous claims for damage compensation poured in from villagers for broken glass and cracked ceilings and walls. The chagrined government had no choice but to pay, for investigating each case would have resulted in even greater liabilities. There were ludicrous claims, however, that the government refused to address, such as the newlyweds who wanted financial compensation for having a baby because Concorde allegedly thwarted their contraception efforts.

The complaints resulted in the government banning supersonic flight over land, a humiliating setback after trying to enhance Concorde's image in the public's eyes—and ears. Ironically, the government's noise tests and attempt to promote good will handed a major victory on a silver platter to the Anti-Concorde Project, whose members couldn't help gloating over the course of events. The die-hard SST opponents saw the ban as an auspicious development in their favor and pursued, with renewed vigor, their relentless goal—the death of Concorde.

The British government soon found itself in another plight. On February 4, 1971, the nation and the aviation world were shocked by an unthinkable tragedy: Rolls-Royce, the preeminent manufacturer of aircraft engines, was bankrupt! Concorde proponents were appalled, for the SST's very heart was her Rolls-Royce power. She would surely die without it. Would the project now come to an ignominious cancellation after so much progress had been achieved?

Thoroughly perturbed but supportive, the British government came to the rescue and poured millions of dollars of public funds into Rolls-Royce to keep the company alive. Concorde was saved.

There was mixed reaction in Europe to the death of the American SST. Concerned that environmentalists would turn their attack on Concorde, Britain cautiously remained silent. On the other hand, France was ebullient and saw America's failure as an unlimited opportunity for Concorde. The French had greater confidence than ever that the world's airlines soon would be purchasing fleets of Anglo-French SSTs. It was plainly obvious with the major competitor out of the running—victory was going to be achieved through a rout!

Already, 16 airlines had placed options for Concorde, including Pan American, TWA, United, American, Eastern, Braniff and Continental from the

United States. The total order now stood at 72 aircraft, but the optimistic Anglo-French sales team was banking on selling at least 200. Indeed, the sky appeared the limit, for even the Federal Aviation Administration in America had forecast a possible market for no less than 500 SSTs across the world's air routes. Not only would France and Britain regain supremacy of the skies, they would recover all that had been invested in Concorde's development and make a sizable profit to boot.

The new leader of France, President Pompidou, was caught up in the euphoria sweeping across his country. To demonstrate his wholehearted support of Concorde, he embarked on a supersonic flight in the craft. While flying 60,000 feet high at twice the speed of sound, he broadcasted to his people: "From the technical point of view, the results have been achieved and the aircraft responds to the most ambitious hopes. From the point of view of money, the government has decided to pursue the project to the end . . . How could anybody imagine that for the first time in history, humanity will recoil before what is a spectacular and peaceful advance?"

France had looked with disdain on the commotion over noise and the environment in Britain. Gallic support of Concorde was virtually unanimous; the Anti-Concorde Project was deemed simply incomprehensible and considered downright unpatriotic. But the French were grateful beyond words to the British government for bailing out Rolls-Royce and saving the SST. It was time to make amends with their counterparts across the channel. At long last, France accepted Britain's application to become a member of the Common Market; the official date of entry was set for January 1, 1973.

In this state of harmony, on September 29, 1972, more than a thousand European statesmen, industrialists and press representatives gathered at Toulouse for a celebration with fanfare suitable for the inauguration of a head of state. Concorde, her grooming session complete and now unequivocally the fastest and most-tested transport in the world, was ready for airline service. The dream had become reality and even the gods smiled on the occasion. They provided a warm, blue sky so clear that the peaks of the Pyrenees could be seen in the distance—as though in admiration of man's victory, achieved through respect for the barriers of nature and not arrogance.

Speeches and music led to Concorde making an appearance before the huge crowd. She was parked front and center for all to behold. Champagne was provided—each bottle displayed a unique Concorde emblem—and toasts were proposed to the supereminent airliner and what she truly represented: The greatest technological *tour de force* in the history of air-transport development.

The Anglo-French sales team had hopes for at least 200 Concorde sales. The FAA forecast a possible market for no less than 500 SSTs. AIR FRANCE PHOTO

Chapter Four:
The World's SST Market Collapses

Now it was time to begin reaping the reward for the successful development of a safe SST through the acquisition of firm orders from the world's airlines. Air France and BOAC (which became British Airways) signed agreements to purchase an initial nine aircraft, but all other carriers remained strangely quiet.

In early January 1973, an elite sales team was dispatched to New York with a critical mission: Sell Concorde to Pan American. The options placed by the foremost American carrier were due to expire at the end of the month; failure to convert them into definite orders would terminate official interest in the aircraft.

It was widely known that Pan American's endorsement of Concorde was the key to an avalanche of orders. Traditionally, the innovative airline set the technological standard to which other airlines kept pace. This was exemplified in 1958, when Pan American launched 707s into service and rendered piston-engine transports obsolete, forcing everyone in the industry also to ditch jugs for jets. The same situation would prevail if the leading airline purchased Concorde: Competing carriers would acquire the SST just to keep abreast of the new mode of air travel lest they fall by the wayside.

The new year of 1973 found Pan American in financial straits, however. A recession, decline of traffic, unfilled seats and resulting poor load factors over the previous three years had forced the airline to borrow a staggering $280 million to keep from foundering. In large part, this havoc was brought on by the introduction of a new airplane—the Boeing 747.

Boeing, with its enormous resources, had developed the "jumbo jet" during its ill-fated attempt to produce an SST. The mammoth 747, with twice the seating capacity of existing transports, made its debut with Pan American on January 22, 1970, for a transatlantic flight from New York to London. The investment for a fleet of 747s and attendant support equipment had been colossal, but there wasn't yet the passenger traffic to fill the capacious interiors of the aircraft. Red ink spilled across balance sheets as a result of the headlong rush to transition into the "wide body" era. The prominent airline was financially strapped.

The sales team met with the consortium of bankers who were bailing out Pan American and defended Concorde's acquisition, despite high initial purchase and operating costs. To support their argument, they produced a survey conducted by an independent New York company. It revealed that most business fliers were willing to pay 40 percent more than subsonic fares to fly Concorde. Such additional revenue not only would offset the costs of fleet procurement and operation, but would generate handsome returns—supersonic service would make money. The bankers were enthusiastic.

Convincing Pan American's executives of Concorde's viability was another matter entirely. Not once was the sales team given the chance to discuss the matter with the airline in an official meeting. Fearful that the absence of formal negotiations signified rejection, the Europeans offered to lease the SST. The offer made no positive impression, however.

On January 31, 1973, Pan American World Airways announced that it did not want Concorde because: ". . . studies indicate that the airplane will be capable of scheduled supersonic service, but since it has significantly less range, less payload, and higher operation costs than are provided by current prospective wide-bodied jets, it will require substantially higher fares than today."

Officials of other airlines had been eagerly awaiting the decision. They remembered only too well how Pan American's early entry into the Jet Age had forced them to discard relatively young propeller-driven Constellations and DC-7s and to spend a fortune on reequipping

British Airways late afternoon departure from London Heathrow Airport of G-BOAE. PHOTO BY ROB FINLAYSON VIA AIRONAUTICA

An Air France Concorde parked in the Qantas engineering area at Sydney Kingsford Smith Airport, Australia, on one of several Round-the-World Concorde flights that the airline has undertaken in 1987 and 1988.
PHOTO BY ROB FINLAYSON VIA AIRONAUTICA

The debut of the Boeing 747 with Pan American on January 22, 1970, found flight personnel in a festive mood, but accountants found balance sheets in the red. Early advent of the wide-body era with lack of passenger traffic forced Pan Am to borrow $280 million to keep afloat — and cooled enthusiasm for acquisition of Concorde.
PAN AMERICAN WORLD AIR-WAYS PHOTO

McDonnell-Douglas entered the wide body age by producing the DC-10. Illustrated is N103AA of American Airlines, first to introduce the type into service — on August 5, 1971.
AMERICAN AIRLINES PHOTO

Lockheed joined the jumbo-jet set with its L-1011 *Tristar*, which entered service on April 26, 1972. Seen here is N31033 in TWA's current paint scheme. TWA PHOTO

with jet transports. It had appeared as though they were going to have to reenact the same scenario, this time substituting SSTs for subsonic aircraft that still had many years of productive life.

Competing carriers also were in financial difficulty, for they had followed Pan American's dash for 747s by scampering for wide-bodied fleets of their own. Now they were experiencing the same problem that plagued the leader: Fleets of jumbo jets were unprofitably flying routes with half their seats empty. A purchase of the SST represented a diversion of funds that nobody had to invest—and wouldn't have for many years.

Almost immediately after Pan American's negative decision, TWA announced its rejection of Concorde. Aside from Air France and British Airways, ultimately every airline in the United States and the world with options on the SST vetoed securing definite orders! That included United Air Lines, which had been very satisfied with the Caravelle. The dream of safe, supersonic commercial flight, which had taken so long to nurture into reality, was repulsed with stunning swiftness.

The reaction in Europe was indescribable. France and Britain were in a state of shock, as though they were parents of a beautiful debutante who had been ignored by society at her acceptance ceremony. Although some blamed the debacle on the economic throes of the airlines, most vociferously denounced the blunt rejection as a conspiracy to kill Concorde and Europe's bid for supremacy of the sky.

Concorde had been dealt a near death blow. None of the $3 billion investment would be recovered now; in fact, millions more would be lost to produce the small order of nine aircraft for Air France and British Airways. Common business sense dictated that the SST was a liability and had no right to live.

The great project, which seemed doomed, was not terminated! Henri Ziegler, a former general in the French Air Force and now president of Aerospatiale, refused to surrender to the cruel circumstances. "Concorde is a good airplane!" he cried out, as he rallied his countrymen to continue with production. Across the Channel, Sir George Edwards, builder of aircraft since the biplane era and now chairman of BAC, demonstrated the same grit and called for his people to "keep their nerve!"

And so these iron men, leaders of French and British industry who embodied the determination that had led to triumph over aeronautical adversity in the quest for safe, supersonic flight, now spurred their nations onward against the greatest of odds. In doing so, they demonstrated to the world that there is something in the human spirit—a blend of faith and fortitude—that can turn defeat into victory.

Sir George Edwards, chairman of BAC, called for Concorde's survival when cancellation of the SST project appeared imminent. BAe PHOTO

Observers around the world were incredulous. Environmentalists were aghast. Executives of airlines that competed with Air France and British Airways were appalled. Concorde still was alive! The news was astounding, almost incomprehensible.

It quickly became clear that the flag carriers of France and Britain were going to mount an all-out offensive with Concorde to capture the lucrative first-class traffic on commercial routes—especially on the precious North Atlantic run. It would be like David taking on Goliath, for the small supersonic fleet was confronting American-made aircraft that now made up 96 percent of the world's market. But studies from the same survey company that previously revealed that business travelers were willing to pay 40 percent more to fly Concorde indicated that supersonic service would capture up to 84 percent of the overseas business market originating in the United States.

Concorde's speed would make up for the small size of the fleet. Her appeal would be further enhanced by a plush interior with superb in-flight service and VIP amenities on the ground—special check-in counters, waiting lounges, personal assistance for connecting flights and hotel arrangements and customs clearance.

This was bad news for executives of competing carriers. Losses from wide-bodies flying around half empty continued in 1973, while a widening recession forecast continued poor load factors into the foreseeable future.

Tragedy struck on June 3, 1973, and caused an uproar among SST opponents. A Russian TU-144, while performing at the Paris Air Show, fell out of control at the top of a steep power climb. The airliner plummeted from 7,000 feet and exploded on impact with the ground. There were immediate claims that supersonic flight wasn't safe, but an investigation failed to determine the exact cause of the accident. No faults could be detected in the aircraft's construction or systems.

Now, SST opponents openly attacked Concorde—in both Britain and the United States. Environmentalists concentrated their criticism on noise and the alleged damage supersonic flight would wreak on the ozone layer in the upper atmosphere. In New York, a bill was introduced that would bar SSTs with the noise level of Concorde from landing at the airport.

Miffed but undaunted, the French and British continued with production of Concorde for airline service. In fact, even though it represented a loss of further funds, they decided to increase the number of aircraft to be built from nine to 16.

Then, an incredible event occurred during the summer of 1973. The people of Dallas and Fort Worth, Texas, who were completing construction of the world's largest airport between their two cities, invited Concorde to attend the airport's inauguration in the fall. Landing permission was granted by the FAA, which, although it had prohibited supersonic flight over the United States, had no qualms about Concorde flying into the country at subsonic speeds. Actually, the FAA was impressed with the engineering and devotion to safety that had gone into the SST's development.

On September 20, 1973, Concorde rocketed across the ocean at Mach 2, slowed down upon entering American air space, landed at Dallas-Fort Worth Airport. A crowd of thousands stood by to greet her, enthralled at the opportunity to see the world's most advanced airliner. Concorde demonstrated her gratefulness for the good will by flying passengers on two supersonic tours over the Gulf of Mexico.

When the festivities in Texas were over, Concorde flew subsonically to Washington, D.C. President Nixon, ever enthusiastic over supersonic travel, bestowed Andre Turcat and Brian Trubshaw with the eminent Harmon Trophy for successfully flying Concorde through her rigorous testing program. Although he fully intended to support the SST's entry into the country for airline service, Watergate soon would close in to prevent him from having any say.

Concorde departed Washington and flew home in a record time of three hours, 33 minutes. France and Britain, instilled with additional confidence as a result of the U.S. tour, investigated modifications that could make the SST more attractive to the airlines. Installation of an extra fuel tank in the fuselage was contemplated to extend the craft's range and enable New York-to-Frankfurt flights, which would enhance her appeal to international carriers.

All optimism came to a sudden halt in October 1973. The Arabs had invoked an oil blockade, sending fuel prices through the roof. Airline executives, confronted with paying 400 percent more for jet kerosene, cut service and increased fares. Any ideas of adding a supersonic gas guzzler to a subsonic fleet were now out of the question.

In February 1974, Prime Minister Heath and his government were removed from office in an election that came during Britain's worst financial crisis since World War II. Harold Wilson and his Labour Party returned to power and immediately inveighed against the SST project's continuance. They cited lack of orders and losses of funds. Worsening Concorde's chances for survival was the release of a damning report in which British Airways estimated that it would lose money on supersonic operations. British newspapers, as a result, predicted that cancellation of the project was imminent.

Employees of BAC, outraged beyond description, fought to save the aircraft they had worked so hard to build. In one of the most emotional episodes in the project's history, BAC workers—from secretaries to top executives—gathered at Filton to remonstrate in a group 11,000 strong. They created an incredible sight, a throng of united humanity thrusting the industry's unequivocal support of Concorde into the spotlight.

To further help their cause, the workers launched a campaign called CASE—Campaign for Action on Supersonic Engineering—and went to Parliament to make known the virtues of the SST project. They argued that Concorde would generate revenue in service, with a break-even load factor of 50 percent; that the funds invested in the project also were funds invested in British technology, creating industrial spin-off of astronomical worth; and that the United States had spent more on research than what Britain had spent on an SST that was a reality.

Additional support was provided by the Confederation of Shipbuilding and Engineering Workers Union, which predicted that Concorde's death would lead to the death of the British aerospace industry and irreparable damage to the nation's position as a technological leader. The union also revealed that the SST project represented only 1/600 of public expenditure—a small fraction of what Britain was spending in areas such as railroads.

To mount more pressure, BAC published a book titled *The Common Sense of Concorde* and distributed it to the public and government officials. Then, hundreds of employees descended on 10 Downing Street in London, calling on the prime minister to spare Concorde's life.

Meanwhile, members of the Anti-Concorde Project struck hard at the SST and urged American environmentalists to force airports in the United States to bar the aircraft from landing. Such restrictions would rob Concorde of the routes most vital for her economic success.

In this war-like atmosphere, President Pompidou, ardent supporter of Concorde, died. The British government waited for the appointment of a new French leader before deciding on the SST's fate; opponents hoped that, at long last, there would be a thaw in the official French position, which hitherto had been zealously in favor of the project. A new leader of France came into power, Giscard d'Estaing, who not only wholeheartedly supported Concorde's survival, but also favored building additional SSTs! Undiminished Gallic enthusiasm was further demonstrated by Air France when it announced that it was confident Concorde would fly filled with business travelers and that super-sonic operations would turn a profit.

The tide was turning in Concorde's favor; British opponents became less vociferous in their demands for the project's cancellation. Then, incredibly, British Airways issued another stunning report in which it alleged that Concorde would not be profitable to operate and that she would harm the financial standing of the airline. BAC was fit to be tied! The aircraft workers looked on this vote of no confidence in the SST as a machination to have the aircraft scuttled. It brought back to the surface bitter memories of how British Airways had tried to refuse purchasing the British-built VC-10 in favor of Boeing airliners. Now, it seemed, the airline didn't want Concorde.

Concorde could withstand no more abuse. She took off for the sky, as if in indignation, and proceeded to perform a series of flights to Rio de Janeiro and back. With American and European businessmen as guests, she demonstrated that she not only could perform airline service with great speed, but also with élan. Impervious to winds, her arrivals were achieved within minutes of the schedule. Turn-arounds were completed free of hassle.

She then flew to Boston, where thousands of Americans jammed Logan Airport to greet her. Heartened by the warm reception, Concorde put her blistering speed into effect and flew to Paris and back to Boston in less time than was taken by a Boeing 747 to complete the same trip one way.

The future of the SST project finally was dependent on the outcome of a meeting between Prime Minister Wilson and President d'Estaing in Paris on July 19,

Henri Ziegler (far right), president of Aerospatiale, defended Concorde and fought for continuing production.
AEROSPATIALE PHOTO

Supersonic airliners under construction. The production team worked hand-in-hand with the engineers—from the concept stage on to completion. The end result was ultraprecise manufacturing and vital weight savings. BAe PHOTO

There were two assembly lines for *Concorde*: one in Toulouse, France and one in Filton, Britain. Illustrated are SST's undergoing final assembly in BAC's Filton plant. BAe PHOTO

British airways
Concorde
Miss Moses Lake
16374
MAINEiacs
AEROS

A recession, the airlines' lack of cash after rushing for 747s and other wide bodies, poor load factors, the fuel crisis of 1973 — truculent circumstances caused the collapse of the world's SST market. Nevertheless, the Concorde program continued for Air France and British Airways. BAe PHOTO

1974. The French position advocated production of the 16 aircraft already slated to be built, plus an additional three aircraft—with modifications to be undertaken to improve performance. The British leader wearily concurred with completing the order of 16 aircraft, but refused to build any more and modify the existing batch unless definite orders were placed by foreign airlines. The Frenchman was disappointed, but did not raise objections for fear of the British pulling out of the project if arguing ensued. So, at last, the matter was settled: 16 SSTs would be built without modifications.

Supersonic proponents were relieved beyond description to see the project saved. But additional orders failed to come from other airlines; thus, the number of production SSTs to be built would not go beyond 16 examples. Aircraft workers would be kept busy for only a couple of years, a far cry from the intoxicating days when it had been envisioned that hundreds of orders would keep production lines going strong for more than 10 years.

In February 1975, Air France and British Airways applied to the United States for approval of additional scheduled flights into the country. They did so via the standard procedure of writing to the assistant administrator for international affairs of the FAA. Each airline requested an amendment of its operating specifications to include Concorde flights into New York (John F. Kennedy International Airport) and Washington, D.C. (Dulles International Airport).

In the past, approvals of amendments requested by foreign air carriers were issued almost automatically. However, there was a high hurdle for the unique Concorde to overcome. The U.S. National Environmental Policy Act of 1969 required that whenever a "major federal action" is considered that might affect the quality of the environment, an evaluation of that action must be undertaken beforehand to determine in detail its impact on the environment. Amending the operations specifications of Air France and British Airways to allow Concorde into the United States for commercial service was construed as a "major federal action."

Part of the process called for an environmental impact statement to be prepared, with its findings exposed to the public for examination and comment. Concorde was the first airplane to be subjected to this formal scrutiny, a fact that didn't sit well with the Europeans, who felt there had been collusion among American air carriers when they all refused to purchase the SST. Now, it seemed, xenophobia was rearing its ugly head to banish the foreign-built aircraft in order to protect American domination of the skies.

The FAA gathered the technical data for the environmental impact statement and published its final findings in November 1975. In turn, Secretary of Transportation William Coleman announced that he would hold a public hearing on the matter on January 5, 1976. From the opinions expressed at the hearing, he was going to decide whether to allow Concorde into the United States for commercial service.

European criticism of the U.S. carriers for cancelling their *Concorde* options was unfounded. Pan American, in particular, far from being a xenophobe, later purchased a huge fleet of European-designed *Airbus* airliners. Seen here is N203PA, *Clipper New York,* an A300B4. PAN AMERICAN WORLD AIRWAYS PHOTO

Chapter Five:
Showdown in Washington, D.C.

The stage was set for the warring factions to clash, after years of combat, in one final conflict. It was to be the most crucial battle, for Concorde needed the routes into the United States to survive in airline service. Everybody knew it. Failure to obtain approval would result in her demise.

The hearing took place in Washington, D.C., with all the drama of a "High Noon" western. Opponents came out in force, including Senator James Buckley of New York, Congressman John Wydler from the 5th District of New York, Congressman Lester Wolff from the 6th District of New York, Congressman Richard Ottinger from the 24th District of New York, Congressman Joseph Fisher from the 10th District of Virginia, Congressman Sidney Yates from the 9th District of Illinois, Congressman John Dingell from the 16th District of Michigan, Congressman James Stanton from the 20th District of Ohio, Raymond Schuller, secretary of transportation of the state of New York, on behalf of Governor Carey; John Shea III, assistant attorney general of the state of New York, on behalf of Attorney General Lefkowitz; the Environmental Protection Agency, the Federal Energy Administration, the Aviation Consumer Action Project, the Airport Operators Council International, the Massachusetts Air Pollution and Noise Abatement Committee, the National Parks and Conservation Association, the Airport Noise Control Committee, the Center for the Study of Noise in Society, the Environmental Defense Fund (backed by the Citizens Against Noise) and the Sierra Club, Friends of the Earth, the National Organization to Insure a Sound Controlled Environment (NOISE), the D.C. Committee for Wilderness, the Fairfax County Park Authority, the Emergency Coalition to Stop the SST, the Community Coalition Council, the Metropolitan Washington Council of Governments, the Arlingtonians for the Preservation of the Potomac Palisades, and additional civic associations.

Joining the attack on Concorde were British representatives from the Anti-Concorde Project, the British Conservation Society and the Heathrow Association for the Control of Aircraft Noise. To the horror of SST supporters, one of the Britishers related the damning estimates made by British Airways during the airline's negative appraisal of the aircraft, revealing that losses of $50 million a year had been predicted. He added, . . . a ban on Concorde would not harm British Airways . . ."

Much of the criticism was directed against noise, with opinions differing on exactly how loud Concorde was: From four to 16 times louder than a 747 to 537 times louder than a DC-10 and 551 times louder than a 707. Many wanted the SST banned because she didn't meet the new noise standards set by Federal Aviation Regulation Part 36; however, 80 percent of the existing airliner fleet didn't meet the new standards, for they were applicable to new aircraft certificated after 1969 and previously type-certificated aircraft that hadn't flown before 1974. In addition, these regulations applied only to subsonics—no noise standards had been implemented for SSTs.

Opponents deplored what they considered excessive fuel consumption; the SST burned more fuel per passenger seat than the 747. But Secretary Coleman revealed that Pan American's 747s were flying with a load factor of 48 percent—more than half empty—and said that this had to be taken into consideration when making a comparison between the two aircraft. Nevertheless, the Environmental Protection Agency and the Federal Energy Administration championed the restrictive 55 mile per hour speed limit and wanted the SST banned as a similar symbolic gesture to the energy-saving campaign that was sweeping the nation.

Fear was expressed over the SST flying through the upper atmosphere and causing skin cancer as a result.

Said cancer was thought to occur through excessive nitrogen pollution destroying the ozone layer. Ironically, no objections were raised over the B-1 bomber project launched by the U.S. government in 1974 with an investment of $2 billion. That project envisioned hundreds of military SSTs plying the same upper atmosphere that 16 Concordes would allegedly be ruining.

The consequences of the Anglo-French SST performing six flights a day into the country were considered threatening to the well-being of humanity. "According to our study, over the next 40 years we can expect an increase of between 46,400 and 70,000 additional cases of non-melanomic skin cancer," stated one opponent. Others expressed similar concern, although without statistics.

Problems were expected to compound as a result: "Depletion of the ozone layer will create a health problem, lead to major changes in food supply, the world temperature and climate changes." This, in turn, led to another upward twist in the spiral: "Even a slight cooling trend would inevitably condemn thousands, perhaps millions, to malnutrition or starvation beyond what is already happening."

Shifting the attack from ozone to pollution, critics made it clear that they looked on Concorde as a filthy beast spewing out tons of foreign matter into the air at an insane rate. The emission of nitrogen oxide, in particular, was alleged to be "32 percent greater than the combined averages of the Boeing 707, Boeing 747 and DC-8 and DC-10." Fear was expressed over a rapid deterioration of air-quality levels.

There were charges made that Concorde was unsafe—"a flying mess," in the words of one congressman. "We know that it lacks safety features and devices that current aircraft already have installed on them, but which are not available on this plane," he added.

Particular criticism was directed to the SST's turning capability immediately after takeoff. It was purported to be unsafe and unacceptable. Range also was thought to be too short for flights into the U.S. "How can you add to our burden even one flight a day arriving in fog or sleet or storm with a critical fifteen-minute fuel supply?" asked an opponent living by an airport.

The leader of a civic group from the New York area near Kennedy Airport lamented the addition of Concorde to the daily traffic. ". . . it is a noisier, it is a dirtier, and it is a more dangerous aircraft than any that are flying in the skies today, using our airport of course," he said. Totally against allowing the craft into the country, he issued what amounted to a threat: "For if this SST is permitted to land, I assure you that this community, right now, is ready, willing and able to close the airport, the entire airport, to all traffic. We feel that we will wreak economic havoc upon the airlines if necessary, because our lives and our comfort depend on it."

The Vietnam War was enigmatically dragged into the battle against the SST. One speaker cited the expense of the Anglo-French supersonic project and then informed everyone that the war had, in comparison, cost the United States $150 billion. Then another assertion was made about the conflict in Southeast Asia: "If there was one lesson from Vietnam, it must be that people cannot tolerate inhuman references to body counts."

What this had to do with the SST was not clear.

In the midst of all the arguments, xenophobia finally revealed itself. It was considered unfair that the foreign-built SST would be "skimming first-class passengers off existing service" and harming the financial standing of American air carriers, especially Pan American and TWA. One individual was, indeed, outspoken: "It is no secret that the Concorde was designed to . . . rival American leadership in aerospace technology, and to re-establish an important European role in that arena . . . It is a venture of rivalry." The message clearly implied that the U.S. government should banish the SST to protect America's supremacy of the skies.

Speaking out in defense of Concorde were, of course, representatives from Air France, British Airways, the manufacturers and the French and British governments. But on hand to lend support was an impressive list of Americans who lauded the Anglo-French SST: Senator Barry Goldwater of Arizona, Congressman Samuel Stratton of the 28th District of New York, Mayor Wes Wise of Dallas, Mayor Clif Overcash of Fort Worth, Ernest Dean of the Dallas/Fort Worth Regional Airport Board, Wayne Whitham of the Highways and Transportation Department of Virginia, Willis Armstrong of the United States Council of the International Chamber of Commerce, Fred Singer from Environmental Sciences of the University of Virginia, Hugh Ellsaesser of the Atmospheric and Geophysical Division of the Lawrence Livermore Laboratory, John Wiley of the Department of Aeronautics and Astronautics of the Massachusetts Institute of Technology, former FAA administrator John Shaffer, Richard Jones of the Virginia Advisory Committee on Aviation, U.S. SST project administrator William Magruder, the British-American Chamber of Commerce, the Citizens for a Better New York, the TRE Corporation, the Committee for Dulles, the Pacific Legal Foundation and the McDonnell-Douglas Corporation. NASA also sent a letter of support.

The Europeans took the floor first and requested permission to fly Concorde into America for airline service on a limited basis, with only four flights a day

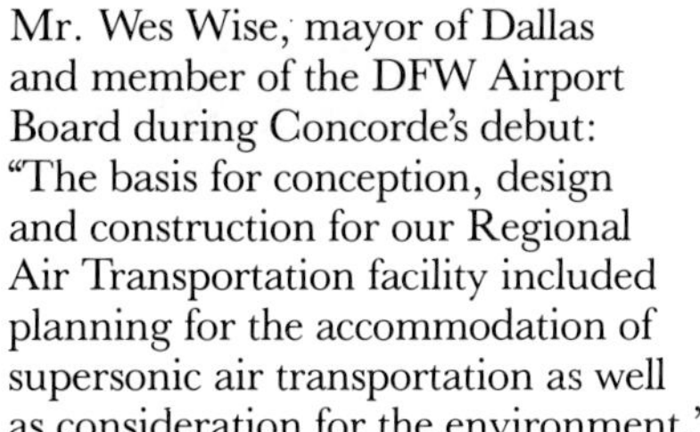

Mr. Wes Wise, mayor of Dallas and member of the DFW Airport Board during Concorde's debut: "The basis for conception, design and construction for our Regional Air Transportation facility included planning for the accommodation of supersonic air transportation as well as consideration for the environment."

Mr. Ernest Dean, executive director of the Dallas/Fort Worth International Airport during Concorde's fight for survival . . . "Just as DFW is the airport of the future, we believe that certainly the Concorde is an aircraft of the future, and the Concorde represents a new form of commercial technology which should at least be given the chance to prove itself."
DFW INT. AIRPORT PHOTO

Mr. Clif Overcash, mayor of Fort Worth when Concorde was a major issue, and member of the DFW Regional Airport board. ". . . Concorde is a tremendous achievement. It is 'state-of-the-art' in commercial aviation today. . . . It implies a harnessing of national will and resources by France and Great Britain of which they may just be proud. . . . I feel that supersonic aviation is an important development in communications between nations."

into Kennedy Airport in New York and two a day into Dulles Airport in Virginia, with no operations before 7 a.m. and after 10 p.m. They affirmed that the request was of vital importance: "The objectives of Concorde cannot be reached unless we fly commercially into the United States."

They defended their SST as a clean aircraft and documented their opinion with studies conducted by the Climatic Impact Assessment Program and the National Academy of Sciences: "Neither claims to present conclusive evidence that a decrease in ozone or an increase in skin disease could be attributed to the proposed Concorde flights." They also pointed out that the environmental impact statement indicated "in essence, that the impact of the Concorde on the American environment will be slight and it will be controllable."

They concurred that the SST was noisier than the new wide-body jets with bypass engines, but informed the hearing that she could meet the decibel standards established at Kennedy Airport since her noise level was "broadly comparable" to the 707 and DC-8. They testified that in view of the very few takeoffs and landings,

the "final environmental impact statement shows that the proposed Concorde operations have a negligible impact on the noise exposure forecasts at the two airports."

In addressing safety, proponents noted that the SST had had more than 5,000 hours of flight-testing—more than any other aircraft—and that her turning capability after takeoff and her range had been deemed totally acceptable by the French and British airworthiness authorities and by the FAA. "We are a respected flag carrier, Mr. Secretary, and we would not jeopardize our reputation by operating an unsafe aircraft," said an Air France official. "We would never put into service an aircraft which had not passed all the stringent tests required of it," said British Airways.

A glimpse into the past was provided to remind how Europe had extended a welcome to America's first-generation jet transports for the sake of fostering aviation progress: "During the late 1950s, Pan American and other United States airlines introduced the 707 and DC-8 aircraft into world passenger carrying service. As part of this process, they requested landing rights to, amongst other destinations, London and Paris. In considering the request for these landing rights, the British and French authorities had to take into account the following facts.

"Both the Boeing 707 and Douglas DC-8 were of noble design, with very many features completely untried by previous civil aircraft . . . Both aircraft demanded considerable increases in available runway length for safe operations. Both aircraft were incomparably noisier than any previous civil aircraft . . . at a time when the British were offering for sale the quietest long range airliner, the turboprop Britannia. The design range of both aircraft was considerably less than the non-stop transatlantic. The economics of operation of the aircraft was such as to demand an increased fare, the jet surcharge.

"Against the background, both the British and French authorities took the decision to grant landing rights, and the direct result is the subsonic civil aviation transportation system which we in the Western world enjoy today," said the BAC official representing the 18,000 members of the Campaign for Action on Supersonic Engineering (CASE), and also on behalf of 20,000 members of the Comite de Soutien.

American proponents supported the statements of the Europeans, with Senator Goldwater taking the lead: "If the United States should reject the Concorde on the ground that the aircraft does not meet noise standards which were unknown and unforeseen at the time it was designed, and I should add, which 2,000 existing United States aircraft also cannot meet, I believe this would be an unfair and discriminatory act against foreign-built aircraft."

He also doubted allegations that the SST would harm the atmosphere: ". . . I am convinced that we have never had any proof, no proof at all, that supersonic fight is causing trouble to the ozone . . . There are only sixteen Concordes committed to production. And in the words of Dr. Paul Campbell, a prominent flight surgeon, who spent a career in research aspects of aviation and space medicine, and I quote: 'One Concorde will have as much environmental impact as a speck of dust in the ocean. Sixteen Concordes like sixteen specks.'"

Scientific people presented testimony that indicated the SST hardly was a threat to the environment. A Professor Singer from the University of Virginia offered interesting insight into the matter: ". . . human influences on stratospheric ozone far exceed those of the Concorde program. For example, the effects of freon gas used in aerosol spray cans and refrigerators should by now have caused a one percent decrease in ozone . . . Other human influences are methane, carbon tetrachloride, and, most important, nitrogen fertilizer used in agriculture. These effects . . . are about ten times as large as the Concorde effect . . . To sum up, as far as the Concorde program is concerned, stratospheric ozone is a non-problem, and so is skin cancer."

Mr. Ellsaesser, a physicist from the Lawrence Livermore Laboratory, cast further doubt on the claims of SST opponents that the ozone layer would be destroyed. He cited examples in which "the ozone data appear to be at variance with the theory of catalytic destruction of ozone by oxides of nitrogen."

At one point in the proceedings, a proponent representing a group of business people concerned with the normal growth of Dulles Airport felt that the attacks on the SST were blown out of proportion: "Many of us were at Dulles on September 26th, 1973, when the Concorde took off for Europe . . ." he said. "All agreed that the noise factor was nothing like the environmentalists claimed. In fact, the noise is slightly more than a DC-8 taking off."

He then expressed indignation at what he considered "a false type of publicity . . . promoted by the environmentalists to stir up citizens who are unaware of the truth."

Secretary Coleman interrupted him: "Wait a minute. Sir, I really do not think it would help now. Let's not kid ourselves. The environmentalists on occasion have alerted this country to problems that we ignored for many years."

Another proponent, Mr. Magruder, former administrator of the ill-fated U.S. SST project, addressed the subject of engineering. His testimony was most supportive: "It is my opinion, after 30 years of flight testing, designing and engineering airplanes, that this could be the safest commercial airplane ever built."

Additional praise for the SST was provided by the representative from the McDonnell-Douglas Corporation. "The present Concorde is a remarkable technical achievement and meets the goals of the designers established for it 11 years and $3 billion ago," he asserted. He called for fair treatment for the Anglo-French aircraft: "Our industry has benefited for over 30 years from non-discriminatory practices overseas. It would seem out of character now to destroy the product opportunities of other countries when they have dedicated so much of their natural resources."

The hearing was adjourned after more than seven hours of testimony. Secretary Coleman promised to reach a decision on the controversial Concorde within 30 days—a decision, he said, that "will not be an easy one to make."

During the waiting period, Air France and British Airways, having received their first production SSTs, launched Concorde into scheduled airline service on January 21, 1976. The French Flag carrier commenced flights from Paris to Rio de Janeiro via Dakar; the British from London to Bahrain. A new epoch of air transportation had begun.

At the same time, the French and British held their breath while waiting for the decision on landing rights in the United States. The fate of Concorde now rested solely on the judgment of one American; all that could be done had been done by the Europeans.

On February 4, 1976, Mr. Coleman presented his decision on Concorde in a voluminous document. It was a masterpiece outlining the arguments, both for and against, with logic and comprehension:

"After careful deliberation, I have decided for the reasons set forth below to permit British Airways and Air France to conduct limited scheduled commercial flights into the United States for a trial period not to exceed 16 months under limitations and restrictions set forth below . . .

"This decision involves environmental, technological, and international considerations that are as complex as they are controversial, and do not lend themselves to easy or graceful evaluation, let alone comparison. I shall nonetheless attempt in some detail to explain my evaluation of the most significant issues . . .

"*Air Quality*: Although there are not currently any emission standards applicable to supersonic transports, the effect of a proposed federal action on air quality is an important consideration . . . The Environmental Impact Statement extensively analyzes the effect of the proposed Concorde flights on ambient air quality, taking into consideration the number of flights, time of day, and wind conditions. The analysis clearly shows that the effect of the proposed level of operations on the air quality is negligible . . . I have therefore concluded, in light of the relatively small number of flights, that the air quality impact of the Concorde is not significant for the purposes of this decision.

"*Energy Impact*: The statistics in the Environmental Impact Statement indicate that the Concorde is less fuel efficient than existing subsonic jets. In absolute terms, the Concorde consumes 20,857 gallons of fuel in a 3,000-nautical-mile trip—about 50 percent more than any of the competing subsonics except the Boeing 747, which, of course, can carry over three times as many passengers as the Concorde. Because of its relatively small capacity, the Concorde also consumes more fuel per passenger mile than the subsonic jets, including the Boeing 747. Flying full, the Concorde consumes .063 gallons of fuel per passenger mile—the equivalent of about 16 miles per gallon per passenger. The Boeing 707-300, the least efficient subsonic jet among those

with transatlantic range, also flying full consumes .030 gallons of fuel per passenger mile—the equivalent of about 33 miles per gallon per passenger . . .

"The significance of these figures is not immediately apparent, however, and will not be until the Concorde is operated in commercial service on the proposed routes, since the actual load factors of the aircraft affect the fuel-efficiency comparisons. If, for example, the Boeing 747 regularly flew at a load factor of around 30 percent—that is, with a number of passengers about equal to the maximum number that a Concorde can carry—and if the Concorde, charging compensatory fares, flew fully loaded, then the Concorde would be more fuel efficient than the Boeing 747 in actual transatlantic operations. I realize, of course, that the actual load factor for Boeing 747s is greater than 30 percent, and that the Concorde with a compensatory fare is most unlikely to operate with a 100 percent load factor . . . But my general point is still valid: The comparative fuel efficiency of the aircraft, measured in terms of the amount of fuel consumed per passenger mile, will lend itself to accurate determination only when we know actual load factors . . . Thus, the market will ultimately be the best judge of relative fuel efficiency . . .

"The Federal Energy Administration and certain participants at the public hearing have argued, however, that I should ban the Concorde as a symbolic gesture . . . But it would be unfair to single out the Concorde as an appropriate symbol of United States fuel-conservation policy. We did not build it. We will not pay its fuel bills. It would border on hypocrisy to choose the Concorde as the place to set an example, while ignoring the relative inefficiency of private jets, cabin cruisers or an assortment of energy-profligates of American manufacture . . .

"*Stratospheric Impact*: The impact of the Concorde's stratospheric emissions on the environment at the surface of the earth has perhaps raised more questions than any other facet of the impact of the Concorde . . .

"The ozone layer is a thin concentration of oxygen molecules, each formed from a combination of three oxygen atoms, which permeates the stratosphere and screens out a portion of the sun's ultraviolet radiation which would otherwise reach the surface of the earth . . . The thickness of the ozone layer at the equator is less than half that at the polar latitudes . . .

"The Environmental Impact Statement, using data from the Climatic Impact Assessment Program report, calculates that the six flights currently under consideration would reduce the density of the ozone layer by about .04 percent, and this reduction would result in an increase in ultraviolet exposure of about .08 percent. The effect of this ultraviolet flux on the incidence of skin cancer is, of course, time dependent—only a prolonged exposure would significantly increase the risk of cancer—but if these calculations are correct, over a 30-year period of continuous operations, the proposed Concorde flights could add an average of around 200 new cases of non-melanomic skin cancer to the approximately 250,000 already experienced per year in the United States.

"There are, of course, substantial uncertainties in this theory . . . Indeed, during the 10 years from 1960 to 1970, scientists monitoring the stratosphere at one location recorded a net increase in the total amount of ozone, while the theoretical models predicted that the injections of pollutants from nuclear explosions in the atmosphere and from elsewhere should have had the opposite effect . . .

"The decision can be put further into perspective by an examination of the relative significance of the ozone reduction caused by SSTs as compared with other sources of increased ultraviolet radiation. The CIAP study found more than 30 possible causes of observable ultraviolet flux on the ground, including cosmic radiation, volcanic activity, accumulation of fluorocarbons from aerosol sprays, subsonic aircraft, nitrogen fertilizer, and nuclear testing in the atmosphere. Fluorocarbons from aerosol sprays and refrigerants are estimated to cause from 12 to 50 times as much ozone reduction as would the Concorde.

"In spite of these statistics, the Consumer Product Safety Commission recently denied petitions to ban these aerosol sprays, based largely on data similar to that before me. And, indeed, the United States harbors a number of sources of stratospheric pollution which have the potential for causing far greater increases in the rate of non-fatal skin cancer than the proposed six Concorde flights to the United States, or for that matter than a fleet of 40 Concordes operating worldwide, yet this country has so far failed to act to ban these sources of emissions into the stratosphere. To protect United States interests on the basis of a conclusion that the evidence is insufficient, and yet to ban the Concorde and thereby threaten the enormous European investment in capital and prestige, would, perhaps justifiably, be perceived by our allies in Europe as discriminatory. Such discrepant treatment would be justifiable only to ward off a substantial and immediate danger of harm, and the danger posed by these flights does not, in my opinion, fall into this category.

"*Noise*: I come, finally, to the Concorde's noise impact, the most significant accurately ascertainable effect of the aircraft. Noise is the factor which has generated by far the greater part of the controversy surrounding the British and French request, and it must be this fac-

tor, if any, that compels me to restrict or ban Concorde flights to either of the two airports . . .

"The Concorde is noisier than the subsonic jets, particularly at takeoff, although the limited flights under consideration will have a negligible impact on total aircraft noise exposure at Dulles and only a marginally incremental impact at JFK. But at a time of heightened environmental consciousness, any additional noise is a serious adverse consequence. The unique characteristics of Concorde noise and the publicity that has surrounded its advent may best be evaluated through a controlled demonstration period of sufficient length to enable an assessment, after the initial publicity has subsided, of community reaction to Concorde noise . . .

"Noise is an objective experience; people do not agree on how objectionable a given sound may be. Noise is a great irritation to some, but of little to others. People may react more intensely to Concorde noise than they would to another new and equally loud source, merely because of the controversy which has surrounded the introduction of that aircraft . . .

"The EIS indicates to me that the marginal impact of six additional flights would be small. Given the subjective nature of human response to noise, however, I must conclude that if any flights at all are justified—that is, if there is sufficient affirmative reason for permitting Concorde flights that we are willing to suffer some environmental effect—those flights should be authorized only on a temporary basis, in order to permit a more intelligent and responsible decision to be made at some point in the future, after we have collected information on the subjective response to Concorde during actual operations.

"*Concorde's Benefits*: . . . The Concorde will significantly increase the speed with which travelers can cross the Atlantic. This will facilitate international commerce and trade and cultural exchange. In addition, the extension to the Concorde of the opportunity to prove itself will be an expression of international cooperation and good will between the United States and two of our closest allies with whom we share a substantial cultural heritage. It will help assure our allies that we seek to act without discrimination and fairly and equitably in our economic relations with them. It will be an important reaffirmation of the mutual reciprocity that has enabled the United States to benefit so substantially from the export of its aeronautical products for the past 30 years . . .

"These and several other considerations have helped impel me toward the conclusion that a demonstration period is appropriate. One is history, which teaches that the benefits of new technologies are not always readily apparent at the time they are introduced. At various points in our history, if we had not acted with foresight we would never have seen the development of the steamboat, the railroad, or the automobile. The advent of the jets was accompanied by substantial debate and public apprehension. Any new technology brings with it a certain degree of risk and a substantial amount of public concern. But the technological innovation that requires tolerance at its inception also enables control of its environmental consequences. I have enough confidence in this nation's environmental commitment and in the objective judgment of the market place to be sure that if the SST does, in fact, become the aircraft of the future, it will only be because man will have developed the technology to meet environmental standards and to enable the SST to compete in the market place effectively. But if we bar the Concorde completely, we may well be condemning for all time or delaying for decades what might be a very significant technological advance for mankind . . .

"A 16-month demonstration period will be sufficiently long to evaluate the variables of seasonality, to help justify the initial commercial investment, to test consumer and community response, and to provide both the European governments and the United States private industry the opportunity to consider whether the development of cleaner, quieter, more fuel-efficient SST technology is a sound capital investment . . .

"Thus, I have concluded that the benefits of an environmentally sound, commercially viable SST would be substantial. I am also convinced that we do not yet have sufficient information upon which to make a judgment about whether such an aircraft could be developed. Given the substantial effort by the French and British to initiate this technology, and the fact that United States participation may well be essential to the commercial success of the SST, I believe this demonstration is needed to determine whether a commitment to this new technology should be embraced."

Mr. William Coleman, secretary of transportation in 1976, presided over the Concorde hearings and embodied the American sense of fair play. His historic decision, one of the most important in aviation history, paved the way for a new future of international trade, commerce and cultural exchange.
DEPT. OF TRANSPORTATION PHOTO

Bill Magruder—test pilot, engineer, administrator—led Lockheed's bid for the SST competition and favored the stationary wing over the swing-wing concept. Later appointed head of the ill-fated U.S. SST project with its Boeing design, he stepped forward to support Concorde's entry into the USA: "This could be the safest airplane ever built. The Concorde is the first commercial airplane that has not had a serious incident during the flight test program and it has flown more than three times the flight tests of anything ever done with jets in this country. This is quite a safety tribute."
DEPT. OF TRANSPORTATION PHOTO

Concorde in flight. G-BOAA was the 6th production machine. "British Airways" was shortened to "British" in 1980, but the full name was retained in the subsequent paint scheme. BRITISH AIRWAYS PHOTO

Chapter Six:
Tribulations and Success

The historic decision was heard around the world—Concorde was allowed into the United States for airline service! Of course, her future beyond the 16-month trial period would be contingent on performing a successful demonstration, but at last the opportunity had been provided for the SST to prove herself.

Air France and British Airways prepared to inaugurate supersonic service from Paris and London to New York and Washington, with May 1976 the target date. The routes to New York were the most important, for they were heavily traveled and rich in business and first-class passengers. But Concorde also was eager to fly into New York for another reason: It would show everyone that she had arrived as a bona fide member of the air-transportation fraternity. For nobody had made it until they had made it in the mecca of the world, the Big Apple.

In March 1976, however, Concorde was dealt a crushing blow that left her gravely wounded: She was barred from Kennedy International Airport! The French and British were horrified at the injury, for the SST depended on New York and would surely die without it.

Environmental groups in the New York area had staged vociferous protests to keep Concorde away. Increased noise from an airport that already had 1,000 aircraft movements a day was the main reason. The proprietor of the airport, the Port Authority of New York and New Jersey, fearing a deluge of lawsuits from irate residents, had reacted by passing resolutions banning the SST to avert liability.

The action raised a serious question over the power of the federal government: Did the Coleman decision preempt this local ban? When Air France and British Airways sought an injunction and a judgment on the Port Authority's right to prohibit Concorde operations, the matter became bogged down in litigation.

In the meantime, Air France and British Airways went ahead with plans to commence service into Washington. On May 24, 1976, SSTs from each airline, filled with passengers, took off from Paris and London and tore across the Atlantic. In anticipation of their arrival—just like thousands of Europeans had gathered to greet Lindbergh 49 years earlier—thousands of Americans jammed the airport and backed up traffic for miles. The sleek airliners appeared over the horizon, landed within minutes of each other, and parked on the ramp nose to nose. The crowd roared—here was history, the first scheduled supersonic service to the United States.

The press also was caught up in the emotion of the event. "THOUSANDS CHEER ARRIVAL OF CONCORDES," read headlines from the *Washington Post*. "FOR CONCORDE, AN AIR OF HISTORY," reported *Newsday*.

Not everyone was enthralled. "Whether it's an economic achievement remains to be seen," a Boeing official from Seattle was reported as saying in the *Newsday* article. "Rather than go for a supersonic, we are trying to develop a very quiet and very fuel-efficient aircraft," said Boeing's Washington representative.

But *Newsday* added a few words of its own: "Of course, there is no shortage of skeptics in the United States when it comes to aeronautical development, which is one of the reasons many achievements in the field have been recorded in Europe."

SST opponents also were on hand—to monitor the aircraft's noise. But after watching the sleek Concordes land and taxi to the ramp, criticism was muted. "It is," conceded John Hellegers to *Newsday*, "a visually attractive nuisance." Previously, he had provided testimony against allowing the SST into the country on behalf of the Environmental Defense Fund.

Newsday also reported that "FAA data indicated one

On May 24, 1976, Air France and British Airways inaugurated supersonic service from Paris and London, respectively, to Washington, D.C. Illustrated is the French carrier's Concorde stopping on the ramp at Dulles Airport; the British SST landed moments later. AIR FRANCE PHOTO

Start of a new epoch: supersonic airline service was inaugurated on January 21, 1976. Air France commenced Concorde flights from Paris to Rio de Janeiro via Dakar; British Airways from London to Bahrain. BAe PHOTO

Unique publicity picture demonstrated just how much progress had been made in transportation.
BRITISH AIRWAYS PHOTO

Publicity photo depicting Concorde's ancestry — the Montgolfier Brothers' hot-air balloon. AIR FRANCE PHOTO

arriving Pan American 707 was louder than the Concordes" . . . and that "many of the more than 4,000 persons jamming observation decks, parking lots and entrance ramps to the airport agreed they were not as loud as they had expected."

Indeed, Concorde's actual noise was quite a contrast to the previously made allegation that she would sound "like one hundred subway trains screeching at one time."

Exasperating for the Europeans, the fight for New York City landing rights continued into 1977. Finally, in May of that year, Judge Milton Pollack of the Federal District Court ruled that the ban was illegal because it was subordinate to Mr. Coleman's federal decision. The Port Authority appealed and the appeal was upheld, with Judge Pollack directed by the appeals court to determine if the ban was discriminatory.

In August 1977, Judge Pollack ruled again that the ban was illegal, for the SST had not been given the chance to prove that she was environmentally acceptable at the airport. Once again, the Port Authority appealed, but this time the appeal was rejected and the airport was ordered to end the ban. The Port Authority sought to appeal before the United States Supreme Court, but the highest court of the land refused to reverse the lower ruling. In October 1977, the ban finally was lifted.

Concorde immediately began proving flights to New York and commenced scheduled service in November. Her takeoff noise was below the 112-decibel limit established by the airport.

With Concorde's legal battle behind her, she embarked on a routine of flying passengers at Mach 2 speed. The calamity dissipated as she gradually gained acceptance by residents of the New York City area. Her actual noise level was well below what many people had been led to believe, for, unknown to the general populace, the Anglo-French manufacturers had addressed the noise issue with an expenditure of an additional $100 million. The SST's wing was modified to enhance low-speed handling at the expense of supersonic efficiency, which allowed her to take off and reduce power to two-thirds thrust for noise-abatement procedures. Suppression was obtained: She now was no louder than a stretch DC-8.

It would have seemed by now that Concorde would finally be left alone to do what she wanted to do—fly. But Hollywood had to cash in on the SST controversy by producing a film on the subject. *SST: Disaster in the Sky* was an asinine movie in which an American SST is crippled in flight and then refused permission to land in Paris because it is carrying a highly infectious flu virus. The ridiculous release would have had more credibility had the Marx Brothers played the flight crew.

Concorde went on to perform impressively. Plying routes at Mach 2 speed with utter reliability, she demonstrated that she truly was a technological tour de force. She also proved to be a very clean airplane, putting to rest prophesies of environmental damage. She emitted less nitrogen exhaust than the 747—only five pounds of pollutant over a course of 1,000 miles. In comparison, the average automobile discharged 50 pounds of pollutant over the same distance.

Concorde successfully passed her 16-month trial demonstration. She received approval to continue scheduled service into the United States, with additional airports made available to her. At long last, the SST was accepted as a valid part of the air-transportation system by the authorities.

In the meantime, the Russian SST project had kept pace, with production TU-144s coming off the assembly line. But changes had been made; the production model presented a sharp contrast to the prototype. The engines now were separated into pairs and were placed further out under the wing. The landing gear retracted into the nacelles and retractable foreplanes— "ears"—were attached to the fuselage to improve low-speed maneuvering.

There were no worries about the TU-144 being repudiated by foreign airports because of noise, for the aircraft's Kuznetsov NK-144 turbofan engines were quiet in comparison to turbojets. Indeed, the advent of a quiet SST would be looked on as the most spectacular development in the history of air transportation.

Route-proving trials took place and traversed the broad country via Moscow, Tyumen and Vladivostock. Aeroflot then commenced supersonic service between Moscow and Alma-Ata, a medium-range route of 1,900 miles. Initially, only cargo and mail were carried, but passengers later were added.

Suddenly, nothing more was heard about the Russian SST. Flights reportedly were halted as silence shrouded the status of the project.

The TU-144 had run into unanticipated performance problems because of the turbofan engines. The cores of the low-bypass powerplants were too small to maintain supersonic speed without continuous use of the afterburners. As a result of this constant need for extra thrust, fuel consumption was enormous and severely cut the aircraft's range.

The Russians silently dropped out of the SST race. They had gone for the brass ring—the development of a quiet supersonic transport—but they fell short.

Concorde now was the sole SST in the sky. She was logging thousands of flights at rifle-bullet speed with a punctuality record of over 99 percent. Passenger loads steadily increased. The breathtaking had become routine; the aircraft's supremacy, unassailable.

The SST's systems proved trouble-free. Her fly-by-wire controls operated flawlessly—not once was it necessary to revert to the mechanical back-up. The pivoting mechanism of the "droop nose" presented no problems. The fuel transfer system worked efficiently. Outstanding pressurization reliably provided passengers with a cabin altitude of 5,600 feet—and hence, more oxygen than on other airliners that offered a 7,000-foot level. The autoland system enabled Category 3 landings in poor weather with reduced minimums, while the lowered nose allowed pilots to see twice as many approach lights at 100 feet and one fourth mile visibility than subsonic jets. The Olympus 593 turbojet engines ran virtuously, and even greater reliability was provided with the installment of improved combustion chambers.

Flying qualities in an airline environment and Air Traffic Control (ATC) compatibility proved outstanding. At Mach 2 speeds a full circle could be made within 20 feet of altitude, a maneuver that when performed in an F-4 Phantom, in comparison, required the pilot to be razor sharp. In the subsonic regime, Concorde reacted instantly and positively, as if responding to the pilot's thoughts. The aircraft could blend in smoothly with conventional jet traffic and required no "special handling" or priority treatment. Indeed, her power and maneuverability allowed speedy ascents and high and close approaches with safety, which also reduced noise exposure over the ground by as much as 60 percent. In holding and diverting-to-alternate airport situations, the Anglo-French SST actually had a greater fuel reserve than required by U.S. regulations—9,400 pounds more than American-certificated aircraft needed to carry. Absolute range—not operational range—of Concorde was a full 5,000 statute miles. Thus, the 3,838-statute-mile flight from Paris to Dulles (the longest leg of the SST's routes) was not the problem it had been purported to be.

There was a little-known virtue about Concorde, and that was her ability to better withstand a windshear condition than subsonic airliners. In addition to the pilots having an angle-of-attack indicator and a high-angle-of-attack warning system, there were, in the words of the FAA, other "technical reasons why wind shear should have less effect on Concorde than on currently operating subsonic aircraft. These include the inherently higher stability of a delta-wing configuration in turbulence, lower wing loading, higher acceleration capabilities."

The French and British were fiercely proud of Concorde flying with Air France and British Airways in such reliable fashion. But there was resentment among many Europeans that the other airlines had rebuffed her, and that the dream of hundreds of SSTs plying the world's air routes had been ruined. Many blamed the Americans for the broken dream. They clung to the belief that there had been a conspiracy in the United States to kill Concorde—a belief that had been reinforced by the bitter fight for landing rights.

While there were many Americans who tried to kill Concorde, it shouldn't be forgotten that many other Americans fought for her life as if she were one of their own. Special thanks was due to the resolute Texans—led by Ernest Dean, Clif Overcash and Wes Wise—who stood up for her amid great controversy and who followed through with support all the way to Washington. And Bill Magruder, the former administrator of the U.S. SST project, a true sportsman who expressed not sour grapes, but respect and greatly supportive testimony, for the superb engineering of his rival. And companies like McDonnell-Douglas, the renowned manufacturer of airliners and military aircraft, which requested fair treatment for a foreign competitor. And, of course, the United States government of 1976, represented by the Secretary of Transportation, the Honorable William Coleman, who saved Concorde when her fate rested solely in his hands.

It was a little known fact that Mr. Coleman once was on the board of Pan American World Airways. But he hadn't let that position influence his judgment in any way.

He related his feelings in no uncertain terms: "It was suggested at the January 5 hearing that the Concorde will 'skim the cream' off the transatlantic market, with disastrous consequences for the already troubled American flag carriers, Pan American and TWA. It has also been suggested that a relevant consideration should be the possibility that the admission of the Concorde to the United States would spell the end of the dominance of the world aviation industry by United States manufacturers, who also deserve protection by their government. Without regard to whether those economic arguments might be true, I have refused to consider wither possibility . . ."

Pan American, condemned by many Europeans for not buying Concorde and thus negating an avalanche of orders from other airlines, was not deserving of being used as a particular scapegoat. Far from being a xenophobe, the foremost American air carrier later would transact one of the largest foreign-designed aircraft purchases in history by ordering 44 Airbus airliners and placing options on 47 more.

Finally, there were the thousands of soldiers—common GIs—who left the United States to fight for the

Normandy Cemetery, overlooking Omaha Beach and the English Channel, is the resting place for 9,386 American soldiers, most of whom gave their lives in the battle of D Day on June 6, 1944.

and . . .

Brittany Cemetery, just outside the French village of St. James, contains the graves of 4,410 American soldiers, who were killed in the Normandy and Brittany campaigns of 1944.

All of these Americans — and the thousands of others — who fought for freedom when tyranny ruled Europe are a part of Concorde. They are a part of all France and Britain.

AMERICAN BATTLE MONUMENTS COMMISSION PHOTOS

freedom of France and Britain (and the world) when tyranny and terror ruled Europe. Many of them never returned to their loved ones; they had given their lives to the cause. And many returned maimed. The friendship between the United States, France and Britain is bonded by the blood of American warriors, blood shed so that the French and British could exist as a free people with a free will, so they could dream bold dreams and achieve great achievements—like the development of a supersonic transport. All of these dead Americans and surviving veterans are part of Concorde. They are part of all France and Britain.

It was unfortunate that Concorde's fleet was so small. Indeed, it was the greatest tragedy in air-transportation history that the fastest and most advanced airliner ever built possessed a numerical strength of only 16. But the SST was born at a bad time and became a victim of cruel circumstances over which her Anglo-French parents had no control.

Had not the airlines been so financially strapped because of early purchases of the 747 and sent further into distress by the lack of passengers flying the capacious wide bodies, Concorde could have enjoyed numerous orders and entered service in fleets around

The SST's takeoff noise was lessened by a $100 million modification of the wing, which allowed better low-speed handling with reduced power. BRITISH AIRWAYS PHOTO

the world. Once entrenched in the market, she could have withstood the fuel crisis and gone on to prosper. Supersonic flight could have become the normal mode of air transportation, while the French and British could have recouped sufficient return on their investment to embark on the development of a larger and quieter second-generation SST.

Concorde also would have fared much better had she been born 10 years later. The improved financial standing of the world's airlines would have fostered numerous purchases, especially with the creative financing plans that became available. Fuel prices dropped, too, which would have added to her attractiveness.

Even the uproar caused by the environmentalists was due, in large part, to the state of affairs when Concorde made her debut. Oceans and rivers were dumping grounds for sludge and garbage, while factories pumped unfiltered waste into the air with callous disregard for effects on the environment. The SST entered the scene when people united to fight the despoliation of their land, which had been taking place unabated for too long. Concorde became a symbolic focus for dissent over all pollution problems.

Alas, Concorde entered the world in a time of economic and environmental strife. She ushered in a new era of air travel and deserved better than what she received, but truculent circumstances gave her no quarter. It was as if she were a child born during the Depression, filled with promise and hope but negated by the very surroundings she was brought into.

As Concorde continued on in airline service, her incredibly reliable performance was dampened by economic losses. Critics of old were quick to point out how she wasn't the viable vehicle her manufacturers said she would be, and laughingly labeled her a white elephant. Air France and British Airways took a hard look at their marketing strategy and decided immediate action was needed in order to reverse the lamentable situation.

Lightly traveled routes were curtailed and the frequency of flights on high-density ones was increased.

British Airways' G-BOAF is the last *Concorde* to have been built, the 16h production model.
BRITISH AIRWAYS PHOTO

Interesting scene for SST afficionados: three *Concordes* on the ramp at John F. Kennedy Int. Airport in New York. A fourth has just taken off.
AIR FRANCE PHOTO

Air France decided to concentrate its supersonic service on the Paris-New York route, even though the Paris-Dakar-Rio de Janeiro run was recording a 65.8 percent load factor. British Airways remained committed to its London-New York route, but extended its London-Washington service to Miami.

Both airlines terminated joint-service agreements with other carriers. Air France and British Airways had extended the Washington service to Dallas-Fort Worth, with crews from the ill-fated Braniff Airlines flying the SSTs for the Texas trip. Concorde wasn't allowed to fly supersonically over land, however, so it really was an expensive subsonic service.

British Airways also left the London-Bahrain-Singapore route. It had been necessary to paint one side of the aircraft in Singapore Airlines markings in order to fly the Union Jack at Mach 2 to the Orient. It was a marketing triumph for Singapore Airlines—and that was about all. Concorde was happy to leave for greener pastures.

Air France and British Airways also increased the fare, for supersonic flight was a premium service and now had to be charged accordingly. Speed, after all, was Concorde's *raison d'etre*.

Overnight, the SST became highly profitable for her owners—and load factors increased despite the higher cost. It was realized that the fare should have been raised long ago. Air France was heartened to see Concorde come into her own after years of providing steadfast support. British Airways found new enthusiasm and considered her the pride of the fleet. One BA official remarked that what had once been looked on as a "wanton financial embarrassment" now was a resounding success.

The New York routes were especially profitable, with numerous repeating passengers who had come to love Concorde. High-powered executives and lawyers, with a mix of entertainers and diplomats—most were American and half were flying the SST more than four times a year. A quarter of Air France's passengers were making the supersonic trip eight times annually. One business traveler from New Jersey, Fred Finn, achieved a place in the *Guinness Book of World Records* for having flown Concorde hundreds of times. At last count, he was on his way to 600.

More and more passengers were flying Concorde for one main reason—to save time. For the upper echelon business types, time—even hours—was money. There also was a small but growing number of vacationers who chose to travel at Mach 2 just for the sake of experiencing the fastest speed next to space flight.

In addition to scheduled service, the charter business became highly successful for both Air France and British

Airways. The latter airline also began flying *Queen Elizabeth II* ("QE 2") passengers on Concorde as part of a package deal with the Cunard shipping company.

In 1983 Concorde produced profits of $15 million for Air France and British Airways. By the end of 1985, the figure had doubled. There was also economic spin-off—sales of subsonic tickets had increased, as well. People were choosing to fly the flag carriers of France and Britain because they were the technological leaders of the industry.

Topping it all was Concorde's flawless performance. In 1986 she entered her tenth year of airline service with a perfect safety record. She had carried nearly 1.5 million passengers while logging more supersonic hours than all of the air forces in the free world combined.

Concorde had proven herself. The fastest airliner in the world now was a valid and viable part of the air-transportation system. For both her owners and passengers, supersonic had become supereminent.

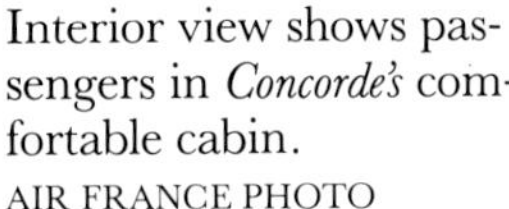

Interior view shows passengers in *Concorde's* comfortable cabin.
AIR FRANCE PHOTO

Concorde is capable of carrying 128 passengers, but Air France and British Airways opted for a first-class layout with spacious seating for 100. BAc PHOTO

The SST now was a valuable financial asset to Air France and British Airways. Still, critics were wont to point out that the reaping of profits never would match what had been invested in the aircraft's development. This was not the fault of Concorde, however. This was the result of the collapse of the supersonic market when other airlines decided against purchasing the Mach-making machine. Concorde, herself, evaluated on her merits as an airliner, was a money-maker for her owners.

When comparing the limited returns with the enormous original investment, there was something else that had to be taken into consideration—the technological spin-off from the SST project. The intensive research conducted by the major contractors and thousands of subcontractors led to dynamic contributions in aviation and other sectors of industry, with long-term economic benefits that were astronomical in worth. The list was impressive and included advanced flight decks and control systems, high-strength and composite materials for aircraft structures, the most resilient glass in the world, high-performance software, advances in aluminum metallurgy, a computer program to define aerodynamic flow around the complete aircraft, innovative laser-manufacturing techniques, digital-controlled machine tools, microswitches able to withstand extreme temperatures, plasma torches for use in the steel industry, specialized oil drilling pipes, reinforced plastics through composition of high-strength fibers, and synthetic oils and fluids.

Research also spawned developments directly helpful to mankind: Aids for the blind and deaf and arthritis sufferers, heart pacemakers and even an artificial heart. Such contributions to society went beyond their economic worth, for the value of aiding and saving lives cannot be measured in terms of money.

Another contribution was made by the Anglo-French SST project. It was the greatest contribution of all and its expansiveness extended through Europe and across the Atlantic to America. Concorde, borne to compete in a world in which cut-throat competition was a way of life, fostered a re-evaluation of nations' ideals and the realization that mankind could be better served by international cooperation.

The French and British had fought to recapture their heritage in aviation from the United States, which, since the advent of the DC-3, had dominated almost the entire world aircraft market. Although Concorde hardly made a dent in that market with her small fleet of 16 production aircraft, her technological superiority propelled France and Britain to the forefront of the industry. The two nations became, unequivocally, the new leaders in air-transport development.

Concorde earned for the Europeans the respect of American industry—and taught it a valuable lesson. Boeing's competing idea of a bigger and better SST proved to be a brash design and ballyhoo ventured as a result of overconfidence from gluttonous consumption of the world's aircraft market. It was so ponderous that it couldn't even take off from the drawing board, while the Anglo-French SST went on to perform Mach 2 flight with a reliability unsurpassed by subsonic jets. Suddenly, American industry had to acknowledge that other countries could build airplanes, too—and supersonic transports, no less. Special attention directed to the allies across the ocean showed a people of substance, people fiercely proud of their heritage and very capable of accomplishing great achievements in the field.

A new outward-looking philosophy transcended old

political barriers. Mr. Coleman embraced a policy of fairness as the secretary of transportation with application to Concorde and the future as well. Long gone, and with good riddance, were the days when emissaries were sent abroad from Washington, D.C., to intervene in foreign aviation affairs. Such invidious actions stained the integrity of American policy, destroyed respect and negated any chance for mutual progress. Thanks to Mr. Coleman—and Concorde for spearheading the issue—a new integrity and a spirit of cooperation has been provided.

The aviation industries of France and Britain went on to grow with a wide range of products—from light utility aircraft to the Ariane space launcher—while other nations in Europe, Asia and South America joined them to participate in the developmental processes. And in this spirit of international alliances, Europe and America reached out and came together, agreeing to span the Atlantic by jointly building an airliner—the Airbus.

Aviation progress had returned full circle to 1785, when a European and an American teamed up to conquer the English Channel by air.

"Air transport is an international industry," reported Airbus Industrie. "Whether it is in the building of aircraft, engines, or the movement of passengers by the airlines, the industry draws its strength from international links."

The Airbus project is a union of great magnitude. A European labor force of 166,000 tradesmen from France, Britain, West Germany, Spain, Holland and Belgium was amalgamated with a legion of 400,000 Americans. This enterprising joint venture, while producing to date airliners comprising only 17 percent of the world's market, nevertheless has, since 1979, generated $3 billion in revenue for the American economy alone.

It is a little known fact that the greatest portion of the value of the Airbus actually is manufactured in the United States—approximately 33 percent. And over a 15-year operational cycle of the airliner, American participation in the project will increase to 45 percent by virtue of using U.S.-manufactured maintenance items, such as engine components, air-conditioning systems, hydraulic pumps, generator-control systems, wheels, brakes, tires, etc. These percentages can be better appreciated by the following translation: Each Airbus produces the equivalent value of a 737 for America's manufacturing industry; over a 15-year cycle, each Airbus will generate more business for the United States than a 727.

There are more than 400 American companies involved in the manufacture of each Airbus. The project is proof that international manufacturing of aircraft can foster trade and a flow of technology between nations, with financial rewards for all involved.

Another notable example of Americans and Europeans working together is provided by McDonnell-Douglas and British Aerospace. In what has become the largest-ever Euro-American military collaborative project, they are sharing development, production and marketing of the Harrier II, a light attack aircraft with vertical and short takeoff and landing capability.

International agreements between the American and British governments are providing procurement of the Harrier II for the U.S. Marine Corps and the Royal Air Force. In addition, an export contract was signed with the Spanish Navy.

Some 25,000 people are involved in the project on both sides of the Atlantic. Distribution of work has been spread to more than 100 subcontracting companies in the United States and Britain. Cooperation also has extended to production of the engine, with Pratt & Whitney becoming the principal subcontractor for Rolls-Royce.

The Harrier II project represents the technological, industrial and commercial resources of two major corporations from two nations. Each provides outstanding achievement in the design and manufacture of civil and military aircraft. McDonnell-Douglas and British Aerospace both are recognized leaders in the aviation industry—together, the transatlantic partnership is unrivaled.

An additional Anglo-American project currently is under way. Under the leadership of McDonnell-Douglas, British Aerospace and Rolls-Royce of Britain and the Sperry Corporation of the United States are developing a new jet pilot training system for the U.S. Navy. Called "Hawk VTXTS," it is scheduled for introduction into the Naval Air Training Command by 1989.

Transatlantic cooperation also has extended to the development of commuter airliners. Actually, the term "commuter" has become a misnomer, for the short-haul passenger aircraft being developed today are scaled-down versions of the transports flying with the major air carriers. They incorporate the latest technology and are very sophisticated, indeed. The SF 340 is one such aircraft, and it recently was produced by Americans and Swedes with British participation, as well.

Fairchild Industries of the United States and Saab-Scania of Sweden joined forces—they called the venture Saab Fairchild International—to develop and build an economical, 35-passenger, turboprop airliner. The wings, tail surfaces and engine nacelles were provided by Fairchild; the fuselage came from Saab. The engines were made by General Electric of the United States,

Although the *Airbus* is European designed, the greatest portion of its value is manufactured in the United States. Here is one of Pan American's A310-200s.
PAN AMERICAN WORLD AIR-WAYS PHOTO

The *Harrier II* is manufactured and marketed by McDonnell-Douglas and British Aerospace.
McDONNELL DOUGLAS PHOTO

while the landing gear and propellers were produced by AP Precision Hydraulics of Liverpool and Dowty-Rotol of Gloucester, Britain, respectively. Final assembly took place in Sweden, whereupon the aircraft was flown to West Malling in Britain for the interior installation and painting by Metair Aircraft Equipment, LTD.

October 27, 1982, was the roll-out date for the first SF 340. Airline executives, VIPs and press representatives gathered at the Linkoping plant in Sweden for an elaborate ceremony, which was highlighted by the King of Sweden signing his name on the new airliner.

Test flights began in 1983 and on May 30, 1984, Swedish aviation authorities certified the SF 340. On June 29, 1984, an historic occasion took place when the other nine countries in the European Joint Airworthiness Regulations (JAR) Group—Belgium, Britain, Denmark, Finland, France, the Netherlands, Norway, Switzerland, West Germany and the United States—met together to issue airworthiness certificates. In October 1984, Australia became the 12th nation to approve the high-tech airliner for commercial service.

To date, 26 airlines in Europe, the United States and Australia are flying the Saab 340. Production is continuing at a steady pace and completion of the 100th model

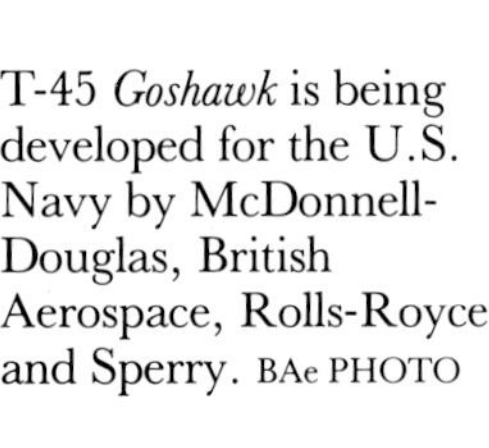

T-45 *Goshawk* is being developed for the U.S. Navy by McDonnell-Douglas, British Aerospace, Rolls-Royce and Sperry. BAe PHOTO

The Saab 340 is presently flying with 26 airlines on five continents. Shown here is Air Midwest's N342AM.
SAAB-SCANIA PHOTO

Acquisition of new F27s by the U.S. Army in December 1985 visibly demonstrated America's recognition of the European aircraft industry. Presently in use with the Golden Knights — the elite parachuting team — the Dutch-built transports are powered by British Rolls-Royce engines and are renowned for their structural integrity and safe performance.
FOKKER AIRCRAFT
U.S.A. PHOTO

is anticipated in 1987.*

The future holds promise for more transatlantic partnerships. One of immense proportion currently is being seriously considered by Lockheed of the United States, British Aerospace of Britain, Aerospatiale of France and Messerschmitt-Bolkow-Blohm (MBB) of West Germany. They are discussing the joint development of an air transport capable of performing with efficiency in both the military and civilian sectors. Called the "FIMA," which is an acronym for Future International Military/Civil Airlifter, the aircraft is envisioned to have a market requiring no less than 2,500 examples. Once the four companies can agree on a final design, the go-ahead for production can be given, and if the project begins soon, it is expected that the FIMA could be ready for flight in the early 1990s.

Transatlantic partnerships have spread to the development and manufacture of aircraft engines, too. CFM International, a joint venture of General Electric of the United States and SNECMA of France, has attained large success in providing powerplants for military and civilian customers. The U.S. Air Force and airlines have retrofitted their Boeing KC-135s and Douglas DC-8-60s, respectively, with the company's CFM56-2 engines. Its CFM56-3 is the only engine available for the hot-selling Boeing 737. Altogether, CFM International has received more than 1,000 orders for its -2 and -3 engines, and it is anticipated that the number will increase to 3,000 in the future, which represents revenue of more than $7 billion. The Franco-American team also has developed the CFM56-5 for upcoming hi-tech airliners in the 150-passenger category, and, already, 160 firm orders and more than 100 options have been received. This amounts to more than $1.4 billion in sales from the new -5 engines alone. The market is expected to expand dramatically in the 1990s.

A most intriguing development in the engine industry has been the recent establishment of International Aero Engines—IAE. Rolls-Royce of Britain, Pratt & Whitney of the United States, Japanese Aero Engines of Japan, MTU of West Germany, Fiat Aviazione of Italy—each company devoted a percentage of its resources to form a consortium. Officially launched on January 1, 1984, IAE currently is at work on a highly efficient engine called the V2500. Slated for certification in 1988 and delivery to customers in 1989, it already has received 169 firm orders and 170 options from six airlines to power their forthcoming medium-size aircraft. This represents $1.6 billion worth of V2500s—quite a lucrative beginning for the young international company.

Another multilateral powerplant project has come into existence. It involves General Electric, Rolls-Royce, SNECMA, MTU, Fiat and Volvo (Sweden). They are refining the development of the CF6-80C2, a very powerful turbofan with low fuel consumption. Air India ordered the model for A310-300 Airbus aircraft; Lufthansa for A300-600 Airbuses; and Thai Airways for A300-600s and 747-300s. Additional sales prospects appear excellent, for the CF6-80C2 is being upgraded to produce even more thrust and will be targeted for 747-400s, advanced 767s and MD-11s.

Concorde paved the way for a future of international projects such as the Airbus, Harrier II, T-45 Goshawk, SF 340, FIMA, CFM International, IAE, CF6-80C2 development. The cumulative worth of these projects alone will far exceed the original $3 billion investment into the Anglo-French SST in a way that couldn't have been imagined 20 years ago. Moreover, the rewards are for both sides of the Atlantic.

Concorde bound the world closer together by her speed and fostered commerce between nations by advancing amity. She has demonstrated the true purpose of the airplane.

GE/SNECMA CFM56-2 engines, manufactured by the American-French concern, converted the DC-8-61 into the DC-8-71. Seen here is Delta Air Lines' N1303L.
DELTA AIR LINES PHOTO

*On November 1, 1985, Fairchild announced a corporate restructuring move (due in large part to the financial burden of the delayed T-46 Air Force trainer program) and that it would cease participation in the production of the SF 340 after the 108th model. Saab-Scania will assume control of the project; however, 70 percent of the aircraft still will be made of American parts.

The British Aerospace 146, an economical jet transport, is powered by four ALF 502 fan-jet engines built by Avco Lycoming Textron of Stratford, Connecticut. Seen here is an example in service with United Express. BAe PHOTO

The Jetstream 31, a short/medium-haul transport built by British Aerospace, is powered by Garrett TPE331 turboprop engines made in the United States. Other American contributions include the avionics and much of the aluminum. Illustrated in flight is a Jetstream serving with an American Eagle carrier. BAe PHOTO

Shorts 360 is a "wide-body" regional airliner powered by Pratt & Whitney PT6-series turboprop engines. The six-bladed propellors are provided by Hartzell from America. Seen here is a 360 in American Eagle colors. (Note: in additional trans-atlantic cooperation, Shorts produces the main landing gear doors for the Boeing 747, wing flaps for the Boeing 757, rudder assemblies for the Boeing 737-300 — and nose cowlings for Rolls-Royce RB211 engines on the 747 and 757.) SHORTS PHOTO

The Fokker 50 is an economical short/medium-haul transport built by the Dutch manufacturer. Power is provided by Pratt & Whitney 100-series turboprop engines. Illustrated is one of the two prototypes. FOKKER USA PHOTO

C-212 (left) and CN-235 (right) are medium-size transports developed by CASA (Construcciones Aeronauticas SA) of Spain and IPTN (Industri Pesawat Terbang Nusantara) of Indonesia. The C-212 is powered by American-made Garrett TPE-331 turboprop engines, and to date over 400 of these aircraft have been sold to operators in 38 countries. The CN-235 utilizes General Electric CT7 turboprop powerplants — and other American-made products, such as Hamilton Standard propellors and air conditioning, Rockwell Collins avionics, and most of the raw materials used in construction. In addition, it is reported that the Northrop Corporation of California has a 13.7% participation in CASA. Certificated in 1986, the CN-235 is selling well. CASA PHOTO

Shorts 330, a local-service transport built by Short Brothers of Northern Ireland, is powered by Pratt & Whitney PT6-series turboprops. Military variants went into service with the U.S. Army and U.S. Air Force. Illustrated is a civilian example flying with Olympic Airways, the Greek national airline. SHORTS PHOTO

The ATR 42, developed by Aerospatiale of France and Aeritalia of Italy, is a hi-tech transport capable of carrying 42-50 passengers. Power is supplied by two PW-120 turboprop engines manufactured by Pratt & Whitney of Canada. The propellors, electronic system and avionics are made in America. Illustrated in flight is an ATR 42 of Pan Am Express.
AEROSPATIALE PHOTO

The British Aerospace ATP, an advanced regional airliner, is powered by Pratt & Whitney 100-series turboprop engines with BAe/Hamilton Standard propellors. The air conditioning packs, avionics and flight control system are also American features.
BAe PHOTO

On June 29, 1984, the Swedish-American twin turboprop aircraft Saab-Fairchild 340, received airworthiness acceptance by the U.S. FAA and nine countries in the European JAR Group. Swedish BCA approval came on May 30. SAAB-SCANIA

1,350 mph at nearly 60,000 feet—faster than the F-16 and at an altitude where military pilots wear pressure suits—Concorde plies the sky with utter reliability while allowing her passengers to enjoy first-class service in short sleeves. AIR FRANCE PHOTO

At 10 in the morning, Captain Michel Butel and his co-pilot and flight engineer—the flight crew of Air France Flight 2 from New York to Paris—enter the company's operations office at Kennedy Airport. Hand-picked to fly Concorde, they are smart of manner and appearance, in pressed uniforms of navy and brass. Social amenities are exchanged with the dispatcher, whereupon everyone gets down to the business at hand. Weather, temperature, flight route, payload, fuel load, V speeds. Planning is thorough, but routine, for another Mach 2 flight across the North Atlantic.

The forecast calls for light rain in Paris, but there isn't much concern, what with the SST's Category 3 landing capability in low visibility. The flight engineer thinks it is actually good news. "I'm glad about that," he says. "My lawn needs to be watered." The others chuckle.

When finished with their preparations, the crew departs the office and heads for the waiting aircraft. Captain Butel leads the way at a brisk pace, with his head high and shoulders back, like a military man from West Point. At the gate he steps outside to walk on the ramp while the others enter the cockpit.

Concorde sits there, sleek and haughty, her whiteness brilliant in morning sunshine. Umbilical cords from ground equipment are hooked up to her fuselage, and the scene is more like that of a rocket launch than an airliner departure.

The pilot walks completely around the craft, surveying her components with the care of a conscientious airman, but also as if regarding the attributes of a beautiful woman. "My mistress," he explains with a smile. After a last glance, he climbs aboard and makes his way to the left seat in the needle nose.

Captain Butel speaks in a voice filled with authority. Deceptively gentle on the ground, he is firm in the aircraft. His sense of responsibility is intense.

The commander of Air France Flight 2 is incredibly experienced. As a teen-ager, he flew 100 missions in the Spitfire during World War II, serving with the "Ile de France" squadron of the Free French Forces. For his action in combat, he was decorated with the Legion d'Honneur and the Croix de Guerre. In 1947, he was hired by Air France, trained in multiengine equipment and procedures, and sent on the line as a co-pilot. He later qualified as a captain and plied the air routes of the world in the DC-3, Languedoc, DC-4, Constellation, 707 and 747 before checking out in Concorde. He has an impressive total of more than 21,000 hours of flying time.

The veteran well remembers his experiences as a fighter pilot. He can vividly recall flying in his maneuverable mount and even the sound of its Rolls-Royce Merlin engine. In fact, it has made him more appreciative of the SST's excellent handling qualities.

"She is a beautiful airplane to fly," he imparts. "You see, Concorde flies like a Spitfire!"

Captain Butel also has strong feelings about those who served in the Allied cause and the preservation of the memories of those who gave their lives. On June 14, 1984, he and his wife, Andree, journeyed from France to McGuire Air Force Base in New Jersey to attend a ceremony in honor of Major John E. Petach Jr., an American fighter pilot killed in combat in 1942, whose surviving mother and wife were now being belatedly presented with his posthumously awarded Distinguished Flying Cross. The French couple joined Wing Commander David Angela from the Royal Air Force of Britain as the European representatives for what proved to be a gathering of eagles. With members of the U.S. Air Force, U.S. Army, U.S. Navy, American Volunteer Group (the Flying Tigers), New Jersey Air National Guard and New Jersey State Police, they paid tribute to their fallen comrade in an emotional event that transcended the passing of the years since World War II.

The passengers of Flight 2 arrive and are welcomed aboard by courteous flight attendants who have been carefully selected by the company for their friendliness, manners, fluency in foreign languages and appearance. Once everyone is buckled into his seat, the flight crew performs its start-up tasks. Engines come to life, support equipment is dragged clear by ground personnel and the SST is taxied slowly to the active runway. Traffic at the airport is light at this hour, so there is no delay in receiving takeoff clearance. Lining up with the center of the runway, full power is applied and the afterburners are lit. Acceleration is comparable to a subsonic airliner initially, but increases rapidly from the force of the tremendous thrust. In less of a takeoff roll than required by a 747, Concorde lifts her nose smartly into the air, defying gravity with a flourish. She climbs out steadily, throttles are brought back for the noise-abatement procedure, then a turn is commenced towards the east for the SST Oceanic Track . . . and Paris.

Flight is limited to subsonic speed in the early stage of the journey, but the climb to conventional-jet altitude is accomplished in just minutes. Past the shores of America, there are no more restrictions and Concorde can put her great speed into effect. Approaching supersonic, the afterburners are lit and the sound barrier is traversed without so much as a quiver. At Mach 1.7 the extra power comes off, but speed continues to build gradually while reaching for the dark blue realm of almost 60,000 feet . . . and Mach 2.

Capt. Michel Butel in the cockpit of Concorde. As a teen-ager during World War Two, he flew Spitfires in combat with the Free French Forces. He began flying for Air France in 1947, and has logged more than 21,000 hours in the air. AUTHOR'S PHOTO

1,350 miles per hour—more than twice as fast as contemporary subsonic airliners. There is no impression of such speed, however, and the flight is glass smooth.

There is little for the flight crew to do but monitor instruments until descent. Back in the cabin, however, the flight attendants are busy pampering their passengers. Earlier, champagne and hors d'oeuvres were served, but now the main course is ready—and the selection of beverages and dinners is exemplary:

Champagne Cuvee LOUISE POMMERY 1979
Chablis "LES FOURCHAUMES" 1983
Chambolle Musigny 1979
Chateau Gruaud Larose 1980

Whisky BELL'S 20 ans d'age
Whisky GRANT'S GLENFIDDISH
Cognac MARTELL cordon bleu
Armagnac JEANNEAU
Eau de vie de framboise THÉO PREISS
Liqueur GRAND MARNIER

Caviar frais
Fresh Caviar

Caille farcie aux truffes et Armagnac
Quail Stuffed with Truffles and Armagnac

Medaillon de veau aux morilles
Medallion of Veal with Morels

Les trois mousselines de petits legumes
Three Baby Vegetables Mousselines

Patisserie
Cake

Fruits frais
Fresh Fruit

Served on an individual table cloth with tableware specially designed for Concorde service by Raymond Loewy of Paris, the meal is sumptuous and the drinks are superb. Adding to the enjoyment is the setting: Luxurious seats inside a cabin renovated by Pierre Gautier-Delaye. It's like being in a first-class French restaurant high above the Earth.

The scene is almost difficult to believe. At a speed greater than the F-16 and at an altitude where military pilots wear pressure suits, civilians in short sleeves are having a feast. It's even more extraordinary because it has become so routine—day in and day out, year after year.

West to east, the flight goes against the clock, and darkness arrives with the stars. Concorde is alone in this upper expanse and high above the subsonic traffic.

The palatable dinner is capped off with coffee and a choice of liqueurs. Music is listened to through headsets.

Nearing the continent of Europe, it becomes time to return to the world. Deceleration and descent begins for a smooth transition down to subsonic speed and the lower depths of the atmosphere. Three and a half hours out of New York, Concorde enters the approach for Charles de Gaulle Airport and slides down an invisible banister. Minutes later the power comes back, main wheels grease the runway, the nose comes down and the SST completes her landing roll. Here is Paris, France.

It is a unique experience. Twice the speed of sound with safety and premier service. No other airliner can come close to matching Concorde. She is incomparable.

With her remarkable structural soundness, Concorde will continue to rule the sky for years to come. No other aircraft will challenge her reign in this century. But someday, in the 21st century, a second-generation SST will be born to exploit the travel advantages pioneered by the French and British. And she will come into the world not as a result of conflicting nations, but of cooperating nations, the product of both sides of the Atlantic working together—to carry on the legacy left behind by Concorde.

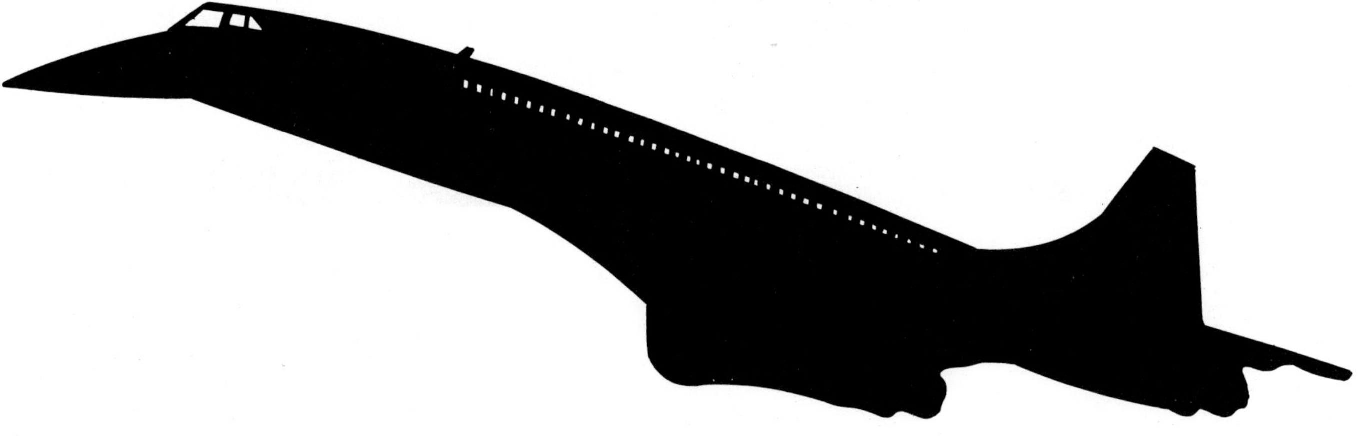

AIR FRANCE

Concorde, the incomparable airliner.
AIR FRANCE PHOTO

Concorde Specifications

Wingspan: 83 feet, 10.4 inches
Wing area: 3,856 square feet
Length: 203 feet, 10.25 inches*
Height: 40 feet, 1.25 inches
Maximum taxi weight: 412,000 pounds
Maximum takeoff weight: 408,000 pounds
Maximum landing weight: 245,000 pounds
Fuel capacity: 210,370 pounds
Empty weight: 174,000 pounds
Zero-fuel weight: 203,000 pounds
Maximum payload: 29,000 pounds
Maximum passenger capacity: 128
Takeoff speed: 225-250 mph
Average takeoff time and distance: 30 seconds and
    1 mile
Maximum subsonic speed over land: Mach 0.93 (at
    29,500 feet)
Supersonic cruising speed over the ocean: Mach 2
    (1,350 mph)
Maximum supersonic speed: Mach 2.02
    (1,370 mph)**
Maximum operating (skin) temperature: 260.6
    degrees F.
Cruising altitude: 50,000-60,000 feet
Average fuel consumed to fly the Atlantic: 160,000
    pounds
Operational range: 4,040 miles (with fuel remaining to
    fly from destination airport to alternate airport
    plus 45 minutes holding)
Absolute range: 5,000 miles
Normal cabin pressure: 5,600 feet
Maximum speed for visor operation: Mach 0.8
Maximum speed for full droop-nose extension:
    310 mph
Maximum landing gear extension speed: 310 mph
Approach to landing speed: 190 mph
Touchdown speed: 170 mph

*In cruising flight, the length is 9 inches longer because of expansion
from heat. (The cabin floor, which doesn't expand, sits on rollers.)
**Fastest transatlantic crossing was achieved by a British Airways
Concorde on December 16, 1978. Time from London to New York 2
hours, 59 minutes, 36 seconds.

The Concorde Fleet

(2 prototypes, 2 pre-production types, 16 production
aircraft)

F-WTSS—Prototype
G-BSST—Prototype
G-AXDN—Pre-production aircraft
F-WTSA—Pre-production aircraft
F-WTSB—1st production aircraft (retained by
    Aerospatiale)
G-BBDG—2nd production aircraft (retained by British
    Aerospace)
F-BTSC—Air France airliner
G-BOAC—British Airways airliner
F-BVFA—Air France airliner
G-BOAA—British Airways airliner
F-BVFB—Air France airliner
G-BOAB—British Airways airliner
F-BVFC—Air France airliner
G-BOAD—British Airways airliner
F-BVFD—Air France airliner
G-BOAE—British Airways airliner
F-BTSD—Air France airliner
G-BOAG—British Airways airliner
F-BVFF—Air France airliner
G-BOAF—British Airways airliner